T0324748

The Geometrical Beauty of Plants

Johan Gielis

The Geometrical Beauty
of Plants

Johan Gielis
Department of Biosciences Engineering
University of Antwerp
Antwerp
Belgium

ISBN 978-94-6239-150-5   ISBN 978-94-6239-151-2   (eBook)
DOI 10.2991/978-94-6239-151-2

Library of Congress Control Number: 2017932536

Printed on acid-free paper

An equation has no meaning for me if it does not express a thought of God.

Ramanujan

"One must study not what is interesting and curious, but what is important and essential".

Pafnuty Lvovich Chebyshev's advice to his students

A special dedication to Walter Liese and Tom Gerats

Preface

In my study of natural shapes, more specifically of bamboo, I started using the superellipses and supercircles of Gabriel Lamé around 1994 to study the shape of certain square bamboos. The first publication was in the Belgian Bamboo Society Newsletter in 1996 followed by a presentation by Prof. Freddy Van Oystaeyen in the same year at a meeting at the University of Louvain organized by the Belgian Plant and Tissue Culture Group and published in the journal *Botanica Scripta Belgica*. Three years later, in 1997, I was able to generalize these curves into what I originally called superformula, as a generalization of supercircles, following joint work with Bert Beirinckx on superellipses. In 1999, I founded a company with the explicit aim to disseminate these ideas in science, technology, and education, with better than expected results.

My first presentation on the more general use of Lamé curves in botany was in 1997 at the Symposium Morphology, Anatomy and Systematics in Leuven, in honor of the great German plant scientist Wilhelm Troll. The symposium was co-organized by the Deutsche Botanische Gesellschaft and by the Botany Department of the University of Louvain. The talk went quite well, and in the closing speech, Erik Smets remarked that it was hoped that I could bring fresh ideas to mathematical botany; the untimely death of the late Aristid Lindenmayer had left a deep gap in that field. On advice of Focko Weberling, one of Troll's students, I was contacted by Springer Verlag that same week to publish a book on my work. In a sense, this book is 20 years overdue, but *pauca sed matura* was Gauss' motto.

In 2003, the first major scientific paper was published in the *American Journal of Botany*, on invitation by the editor in chief, Karl J. Niklas. The title *"A generic geometric transformation which unifies a wide range of natural and abstract shapes"* expresses the gist of the matter. This publication attracted a lot of attention, and I still think it was a good timing and (hoped for but unexpected) strategy for dissemination. In the same year, the English version of my book *Inventing the Circle* was published, two years after the Dutch version in 2001. My journey took me from horticulture and plant biotechnology to geometry. The article *"Universal Natural Shapes"* with Stefan Haesen and Leopold Verstraelen in 2005 introduced the equation into the field of geometers, and also substituted the name superformula by the names *Gielis curves, surfaces, (sub-)manifolds, and transformations.* In geometry, a whole new world unveiled before me. This, along with several publications by others ensured the adoption and absorption of the formula in mathematics, science, and education. Two of my main goals, formulated in 1999, have been or are being realized, along various paths.

Technology was my third, long-term goal, and many papers in science and technology have been published using the formula. In many cases (antennas and nanotechnology, for example), the formula allowed to go beyond the classical and canonical shapes, opening many doors. My own passion about technology is this: No matter what field we consider, I (with many others) think our current technology, no matter how advanced, is essentially a bag-of-tricks, aimed at deception (which is perhaps a major feature of our times and culture). We are working toward new applications following my dream of unifying and simplifying, at the same time appreciating complexity.

In this respect, it is important to note that my background is horticulture and plant biotechnology. I have been involved as researcher and research director in plant research for more than 25 years, and methods were developed in our team for mass propagation of plants, in particular temperate and tropical bamboos, the former for ornamental purposes, the latter for reforestation in the tropics. Over the past years, we have produced over 20 million bamboos that have been distributed and planted worldwide. Bamboo is indeed a multipurpose plant, a beautiful plant for our gardens but providing building materials, food, and much more for the poorest one billion humans on this planet.

Key to this was focused in-depth research using molecular markers and high-throughput determination of plant hormones, but never losing sight of the end goal: plant production. The same procedure of combining science with technology, we use now in the development of antennas, where optimization is an ongoing activity. We are now able to produce very powerful antennas, at costs which could be up to ten times less than existing ones, optimizing margin, while delivering the highest possible quality and efficiency. In all my (scientific and engineering) activities, this combination of wide interests, a generalist (rather than myopic) view, and stamina has always led to remarkable results.

Always keep focused on what you want to achieve. The current book is a combination of such focus, combining wide interests (nature and science) and a generalist (rather than specialist) attitude, inspired by the vision of natural scientists and philosophers from a long gone era. Along the way, I learned many other things, which one cannot learn but by a constant drive and strive to understand. My scientific education continues daily.

I wish to convey my sincere gratitude to my parents and all teachers, botanists, mathematicians, and engineers who played an enormously important role in my personal scientific development and the various developments described in this book: These include my teachers in high school Fred Verstappen (Greek) and Gerard Bodifée (sciences); in my professional horticultural and plant biotechnological life: Pierre Debergh (University of Ghent), Walter Liese (University of Hamburg, Germany), Tom Gerats (Radboud University, The Netherlands), and

Paul Goetghebeur (University of Ghent, Belgium); in mathematics and geometry: Freddy Van Oystaeyen (University of Antwerp), Leopold Verstraelen (University of Louvain, Belgium), Paolo Emilio Ricci (La Sapienza University, Rome), and Ilia Tavkhelidze (Tbilisi State University, Georgia). During the past two decades, fantastic collaborations developed with Bert Beirinckx, with Diego Caratelli (Antenna Company, University of Tomsk, Russia), Yohan Fougerolle (University of Dijon, France), Dishant Pandya (India), and with Shi Peijian, Yulong Ding, and the team at Nanjing Forestry University. I must also thank all collaborators in Genicap Beheer BV and Antenna Company.

Special thanks for their important contributions and advice in various chapters of this book are due to Yohan Fougerolle, Diego Caratelli, Dishant Pandya, Paolo Emilio Ricci, and Ilia Tavkhelidze. The results of the collaborations with all four are fundamental to this book. Many thanks also to Albert Kiefer for graphics to Violet for help on references, corrections and pictures, and to Arjen Sevenster and the Atlantis and Springer teams.

Many thanks to my whole family, past–present–future, for their continuous support. With a most special dedication and many, many thanks to my wife and soul mate Christel, and our children, Violet and Fabian, for making my life complete.

Antwerp, Belgium Johan Gielis
December 2016

About the Book

This book focuses on the origin of the Gielis curves, surfaces, and transformations in the plant sciences. It is shown how these transformations, as a generalization of the Pythagorean theorem, play an essential role in plant morphology and development. New insights show how plants can be understood as developing mathematical equations, which opens the possibility of directly solving analytically any boundary value problem (stress, diffusion, vibration...). This book illustrates how form, development, and evolution of plants unveil as a musical symphony. The reader will gain insight in how the methods are applicable in many diverse scientific and technological fields.

Contents

List of Figures

List of Tables

Part I
Πρότᾰσις—Propositio

Chapter 1
Universal Natural Shapes

The geometrical description of curves and surfaces and the shapes that are derived via Gielis-transformations, describe and determine in a uniform and universal way an enormous diversity of natural shapes.

Leopold Verstraelen

Conic Sections at the Core, Once More

In the book *Inventing the Circle* [1] it was shown how one generic geometric description, a generalization of Pythagoras and Lamé, allows for the description of many natural shapes, illustrated with many pictures and illustrations. The same ideas were published in the *American Journal of Botany* in April 2003 [2]. In 2004 this generic geometric transformation was named Gielis Transformations, giving rise to the class of Gielis curves and surfaces, by the geometer Leopold Verstraelen, who understood that this transformation allows for a broadening of some crucial concepts in geometry. Its actions on classic curves could describe a wide range of shapes in nature [3]: *"The basic shapes of the highly diverse creatures, objects and phenomena, as they are observed by humans, either visually or with the aid of sophisticated apparatus, can essentially, either singular or in combination, be considered as derived from a limited number of special types of geometric figures. From Greek science up to the present this is probably the most important subject of natural philosophy... The geometrical description of curves and surfaces and the shapes that are derived* via *Gielis-transformations, describe and determine in a uniform and universal way an enormous diversity of natural shapes"*.

Because of this wide applicability—the Gielis Formula describes shapes at nano, micro, macro and gigascale—the idea of Universal Natural Shapes was born, providing for a uniform description of natural shapes (Fig. 1.1). T. Philips of the Courant Institute wrote [4]: *"A botanical Kepler awaiting his Newton"*. Obviously, this waiting can take a while. In the meantime we have deepened our understanding of its applicability to study natural shapes and broadening of concepts in geometry and mathematics. This book describes some of these developments, and shows how

© Atlantis Press and the author(s) 2017
J. Gielis, *The Geometrical Beauty of Plants*,
DOI 10.2991/978-94-6239-151-2_1

Fig. 1.1 Universal Natural Shapes. Copyright Johan Gielis/Martin Heigan

it connects various ideas and fields, starting from the principles developed by Ancient Greek mathematicians.

Like *Inventing the Circle* this book is one of ideas, rather than technical. It can be considered as the second part of a trilogy. *Inventing the Circle* introduced the new transformations, and provided many examples of such shapes in nature. In this book we will focus on some of the underlying geometrical aspects based on the subtitle of the first book: *The geometrical beauty of nature*.

We will follow the universal scheme of building arguments as developed by the ancient Greeks [5], the Demonstratio artis geometriae.

Πρότᾰσις or Propositio exposes what needs to be shown, namely that beyond the mere analogy, we can develop a rigorous geometrical and mathematical approach to study natural shapes based on Lamé-Gielis curves and surfaces.

Εκθεσις or Expositio describes what we have to start with. From elementary notions in mathematics that are at work in every field of mathematics and its applications in the natural sciences, we derive Lamé-Gielis curves and surfaces.

Διορισμός: in this Determinatio step we investigate *"whether what is sought is possible or impossible, and how far it is practicable and in how many ways"*. It will be shown how the shapes and their combinations can be easily combined with many existing concepts in geometry and mathematics. This wide applicability applies to a wide variety of natural shapes, Universal Natural Shapes.

Κατασκευή or Constructio: Observations from botany open the door for understanding many key concepts and their connections and allow for analytical solutions of boundary value problems, using classical 19th century methods, thereby broadening various concepts (Gielis is an acronym) and analytical methods. What we can learn from flowers and plants can be applied in virtually all fields of science and technology.

Απόδειξις or Demonstratio: in the final chapters it is demonstrated how Lamé and Gielis curves provide better models for studying various natural shapes, and how they can help understanding evolutionary and developmental aspects in biology, based on geometrical considerations.

Συμπέρασμα: This then leads to the conclusion that many of the suggestions made in the original book and articles in the period 1999–2005 have been validated. Description always precedes understanding the connections between mathematics and nature. Moreover, Lamé-Gielis curves and Gielis transformations have opened the door for many new developments, among others for a geometrical theory of shape and morphogenesis.

Chapter 2
Towards a Geometrical Theory of Morphogenesis

S'il n'est aucune marche tracée pour la recherche de la solution d'un problème, par la simple considération des figures, cette incertitude est peut-être une des grandes richesses de la Géométrie. Il arrive souvent en effet, qu'en suivant certaine route pour aller à la découverte, on rencontre la solution d'un problème différent du proposé, qu'on n'eût sans doute pas trouvé aussi facilement, s'il se fût agi de le résoudre directement. Qui doit-on donc le plus admirer, ou du calcul qui commence ses travaux avec confiance, et finit presque toujours par répondre à la question, ou de la Géométrie qui part sans rien promettre, et revient quelquefois offrir avec la solution du problème, celle de plusieurs autres qu'on ne lui demandait pas, et qu'elle a recueilli sur son passage?

Gabriel Lamé (1818)

Biology Versus Physics

Understanding life is one of the major challenges for science in the 21st century. Despite the exponentially growing mountains of data in the life sciences, the challenge of developing geometrical models, always at the core in eras of scientific progress (Newton, Riemann, Einstein), remains completely open. Marcel Berger wrote explicitly: "*Present models of geometry, even if quite numerous, are not able to answer various essential questions. For example: among all possible configurations of a living organism, describe its trajectory (life) in time*" [6]. We are still far from describing life mathematically, despite the numerous successful applications of mathematics in the life sciences.

Applying mathematics in the life sciences brings hope that what works in physics will also work in biology. In physics, one of the most widely used quotes is *The Unreasonable effectiveness of mathematics in the natural sciences* by Eugene Wigner (1902–1995). A geometrization of physics then seems to be a simpler task than a geometrization of biology. As the Russian mathematician I.M. Gelfand (1913–2009) who had a great interest in biology said: "*There exists yet another phenomenon which is comparable in its inconceivability with the inconceivable*

© Atlantis Press and the author(s) 2017
J. Gielis, *The Geometrical Beauty of Plants*,
DOI 10.2991/978-94-6239-151-2_2

effectiveness of mathematics in physics noted by Wigner—this is the equally inconceivable ineffectiveness of mathematics in biology" [7]. A geometrization of biology, or more generally of nature, based on forms and formation of natural shapes (a geometrical theory of morphogenesis) is both an enormous challenge and a prerequisite for progress in science and the life sciences.

"*Why geometry? If we wish to find the solution of a problem along routes that have not been explored before, comparing geometric shapes is one of the greatest powers of geometry*" [8]. Gabriel Lamé (1795–1860) wrote this in 1818, and he continued: "*What should we admire most? Algebra that starts its search for a solution with confidence and almost always arrives at the response to the question? Or Geometry that starts out without promising anything, and very often delivers a solution to the problem along with solutions to many other problems that were even not asked for, but which can simply be harvested along the way?*" Gabriel Lamé was one of the leading mathematicians in the 19th century, and a dedicated researcher in search of nature's most hidden secrets.

Our goal, and the goal of this book, is to offer a contribution to a geometrical theory of morphogenesis where shape and its formation, the development and evolution of natural shapes, both living and non-living, can be studied in a mathematical way.

A geometrization of biology, or more generally of nature, based on forms and formation of natural shapes is an enormous challenge. René Thom (1923–2002) wrote: "*That we can construct an abstract, purely geometrical theory of morphogenesis, independent of the substrate of forms and the nature of the forces that create them, might seem difficult to believe, especially for the seasoned experimentalist used to working with living matter and always struggling with an elusive reality*" [9].

Such theory would be structurally similar to Newton's geometrization of gravitation. Before Newton a diamond and a piece of black coal were different. After Newton they were still different forms of carbon, but they behaved exactly the same when thrown over the wall. This geometric theory for morphogenesis will not be presented in this book, since it is not yet developed. But we can outline what it might or should look like.

Ideally the geometrical theory of morphogenesis would be comparable to Newtonian theory of gravitation, whereby a uniform description of anisotropy (in the case of planets and comets, the family of conic sections), is followed by a mathematical theory in terms of isotropy. To make this more precise, the first step is the Kepler-Galilei step. Kepler showed that the orbits of planets are not circles but ellipses, and Galilei showed that the trajectory of stones, cannon balls and water jets are parabolas [10]. This provided a uniform description in terms of conic sections. These conic sections had been known for two millennia, and they turned out to be very natural, describing the trajectories of planets and projectiles.

The step that Newton took was to determine the curvature of curves (or trajectories of particles) in terms of local operations, by fitting of the circle to a point on the curve. The radius of the circle is then a measure for the curvature. This idea was suggested already much earlier by Nicolas Oresme, but using analytical methods Newton could achieve this for all planar curves using the second derivative, with wonderful results. Erwin Schrödinger (1887–1961) understood this very well: *"The logical content of Newton's first two laws of motion was to state, that a body moves uniformly in a straight line, ..., and we agree upon calling force its acceleration multiplied by an individual constant. The great achievement was, to concentrate attention on the second derivatives – to suggest that they – not the first or third or fourth, not any other property of motion – ought to be accounted for by the environment"* [11].

Two main lines of development in science may be distinguished. One line focuses on pure description, where conic sections show up. Here we find such names as Apollonius of Perga (who developed the full theory of conics and conic sections), Kepler and Galilei (who applied conic sections to describe trajectories of planets and projectiles) and Gabriel Lamé. The second line focuses on practical computation, using the isotropic space with the Euclidean circle as the unit circle with Archimedes, Ptolemy, Newton and Fourier as most important representatives.

The tendency in science seems to me (1) to describe or understand nature through conic sections, with very compact mathematical descriptions, and (2) to use these insights to develop computational models of nature using isotropic spaces, Euclidean circles, and analytic methods based on series as was done by Newton.

The second step in our project of geometrical theory of morphogenesis would then be the Newton step, generalizing circles and known or new analytic methods based on infinite series. If we can develop a uniform description from within our existing mathematics and geometry, there is less need for "new mathematics", having little or no connection to what we already have.

If we wish to develop analogous steps for a theory of morphogenesis, this would require a uniform description of natural shapes as a first step. This Kepler-like description should be as close as possible to circles and conic sections, and to the Pythagorean Theorem and Euclidean geometry, as almost all scientific developments have followed this path (often in disguise). What will be presented in this book is this first step.

Gielis transformations will be the main ingredient. The original name for these more general transformations was superformula, but the name Gielis curves, surfaces and transformations was preferred by mathematicians, a great honor, no less [12–15]. They are generalizations of Lamé curves and surfaces, named after Gabriel Lamé. They can be considered as a one-step generalization of conic sections.

Commensurability, Symmetry and Euclidean Geometry

With the requirement of a close connection to Euclidean geometry and the Pythagorean Theorem we follow the lines of rational thinking laid down by the ancient Greek mathematicians. The Pythagorean Theorem and Euclidean geometry have been extremely successful, but the legacy of the Greek mathematicians is far richer in ideas than is generally thought. The central ideas relate to commensurability, symmetry and harmony. They are intimately connected and form the basis of this book: *The geometrical beauty of natural shapes.*

The question of *commensurability* is at the very heart of science and mathematics. How does one compare the shapes of flowers and circles? Measuring (μετρεω) is key, but how does one define a common measure or ruler for starfish, squares, cubes, cylinders or a variety of less regular shapes? And how does one describe the transition from a circle to a square, or from an egg to a starfish? Is there a common ruler or compass for such purposes? How does one pass from isotropy—the idea of (the ideal) circle, as the locus of all points at the same distance from a central point—to the anisotropy of natural shapes?

This is the reason why *symmetry* is so central in mathematics. Symmetry (with —μετρια as its root) for the Ancient Greek mathematicians means proportion or right balance and to "symmetrize" (συμμετρεω) is the deliberate act of measuring by comparison, thus making objects commensurable, forming the real basis of mathematics and geometry. Commensurability can come in varieties, the two extremes being *shape* in topology, whereby a torus and a teacup have the same topology, and *measure* in geometry, with well defined ways of measuring. These can be same (or constant) length, same area (equi-areal) or n-volume (*equi-n-areal*), same duration or isochrony (time), or any scalar valued parameter (isobar, isotherm…). The notion of conic sections is based on the same-area-invariant of a rectangle and a square; παραβολη (parabola) means exact fitting, while ellipse and hyperbola are too much and too little respectively, for a perfect fit [5].

Symmetry and commensurability are the reasons why geometry remains at center stage. Geometry, analysis and algebra are the central trinity in mathematics, but Shiing-Shen Chern (1911–2004) wrote: "*While analysis and algebra provide the foundations of mathematics, geometry is at the core*" [16]. There will always be the need for geometers and their geometric sense. André Weil (1906–1998) wrote at the end of the 20th century: "*Whatever the truth of the matter, mathematics in our century would not have made such impressive progress without the geometric sense of Elie Cartan, Heinz Hopf, Chern and a very few more. It seems safe to predict that such men will always be needed if mathematics is to go on as before*" [17].

One of the most defining notions in geometry is that of curvature. Nicolas Oresme's work in the Middle Ages focused on geometry and his representation of shapes by coordinates trying to find a common measure with *curvitas*, curvature based on a circle, as a measure for deviations from straightness [18]. His motive was that clear: For him the measure of phenomena (*De latitudinibus formarum*) should be geometrical. In this sense geometry is the act to make things stand still in a world where everything is constantly moving and changing. This may be one of the reasons why we associate Greek geometrical ideas with rigid geometry. It also carries the idea that Euclidean geometry deals with statics, not dynamics, the language of change. On close inspection however, the very notion of differentiation (*change*) and integration (*quantification of change*) are already encoded in the parabola.

Shiing-Shin Chern clearly expressed this notion: "*Euclidean geometry is the geometrical treatment of the number system*" [19]. The Greek notion of number also included irrational numbers, their original invention. Bartel Leendert Van der Waerden (1903–1996) wrote: *"The reason why the Greeks had to translate the algebraic problems and solutions of the Babylonians into geometrical language, is the discovery of irrationals. The solution of equations in numbers, for the Greek this meant rational numbers, is no longer possible, and the Greeks would not settle for approximate solutions. The geometrical solution is exact and generally applicable"* [20].

The classic inequality $GM < AM$ states that the geometric mean GM is strictly smaller than the arithmetic mean AM. That very same inequality forms the basis for surface theory coupling intrinsic and extrinsic curvature of shapes. In its more general form it is at the basis of most of our theories in physics that evolved in the 19th century starting with Euler and Gauss.

Today Euclidean geometry is less popular. With many new developments in the 19th and 20th century, Euclidean geometry became only one among many possible geometries, linked to specific invariants. Its dominating position in education was considered as a hindrance. *"Euclid must go"*, was stated by Jean Dieudonné (1906–1992) towards new approaches for teaching mathematics in the mid 20th century [21] (although Dieudonné clearly recognized that our current science is built on the foundations that Ancient Greek mathematicians established).

However, Euclidean geometry remains the pillar of our mathematics and our science. Some of the great mathematicians of the 20th century considered Euclidean geometry as very central. René Thom (1923–2002) was, as usual, very outspoken: *"The dilemma posed all scientific explanation is this: magic or geometry. Classical Euclidean geometry can be considered as magic; at the price of minimal distortion of appearances (a point without size, a line without width) the purely formal language of geometry describes adequately the reality of space. We might say, in this sense, that geometry is successful magic"* [9].

A.N. Alexandrov (1912–1999) contributed to mathematics under the slogan: *"Retreat to Euclid"* [22, 23]. For him the pathos of contemporary mathematics is the return to Ancient Greece and also in our view Euclidean geometry remains central. In the case of relativity theory the Euclidean spaces had to be amended to include Lorentz transformations and curved surfaces and manifolds. We will also amend the rigidity of Euclidean geometry with appropriate transformations (in our case Gielis transformations) but it will leave the core untouched. The title *Geometrical beauty of Plants* refers to these ancient roots, whereby beauty refers to the harmony that was so dear to the Greeks.

Mathematical Theories Versus Theories of Everything

When we consider a mathematical or geometrical theory of morphogenesis, we will deal with mathematical propositions or "quantities of M-type" (axioms, assumptions, definitions and theorems). In general, symbols are used to form formulas and equations, but many can be stated in words as well. Ancient Greek geometers stated their propositions in words and in my opinion this is an excellent approach to improve understanding. The Greek method of thinking in mathematics and general logic or reasoning followed the same rules and in a certain way, they did not lack symbols, but they *mastered language*. The rules of reasoning, according to Proclus are the following:

Propositio or πρότᾱσις
Expositio or ἔκθεσις
Determinatio or διορισμός
Constructio or κατασκευή
Demonstratio or ἀπόδειξις
Conclusio or συμπέρασμα

In the Determinatio step one investigates "*whether what is sought is possible or impossible, and how far it is practicable and in how many ways*". If we apply a certain method, it would be good to investigate, how it fits into existing fields or can open new doors, in theory or practice. One important aspect is the existence of a solution. From the 19th century onwards, such existence proofs became crucial in mathematics, but for the Greek this was an absolutely normal step.

The mathematical quantities (M-type) are then compared to observations and data from the fields of physics and biology, from chemistry or astronomy (O-type, O standing for observations), to determine possible identification between the two types. In his book on general relativity theory George Yuri Rainich (1925–1968) expressed this very precisely [24]: "*If these relations coincide (within experimental errors) with relations which appear to be accepted by physicists and astronomers, we say that these last relations, or the experimental and observational facts expressed by them, are explained on the basis of the mathematical theory: in some cases when the relations stated in the formulas have not been observed prior to their derivation, but are observed after, we speak of predictions that have been verified. In both cases we speak of confirmation of the theory of observation.*

This in the author's opinion is the nature of application of mathematics to physics. It is not asserted that there always exists a theory explaining all the facts, and it is not denied that there may exist at the same time two theories which explain (within experimental errors) the known facts equally well. Thus neither uniqueness, nor existence is claimed for a mathematical theory of physical phenomena. In this respect it differs from the concept of truth held by some people. The question whether there is such a thing as truth – questions of the type "What does actually happen?" questions about "physical reality" – we do not consider in this book."

In this book we do exactly the same, namely describing the transformations and see in how many ways they may be applied to the study of nature, without pretending, even not for the geometrical theory of morphogenesis, to work on a Theory of Everything. As Leonard Euler taught us: "*Although to penetrate into the intimate mysteries of nature and thence to learn the true causes of phenomena is not allowed to us, nevertheless it can happen that a certain fictive hypothesis may suffice for explaining many phenomena*" [25].

When we refer to the notion of explaining, it means not only that we can describe shapes up to a desired precision, but that the very mathematical theory gives us the insight in why shapes are the way they are, or why planets follow their elliptic trajectories. For now we refer to this as natural curvature conditions, related to inequalities between intrinsic and extrinsic characteristics of the shapes. As an example, a hanging chain assumes the shape of a catenary realizing equality in the inequality between GM and AM for two functions (e^x and e^{-x}) in the lowest part of the catenary, since both are 1. The arithmetic mean is the hyperbolic cosine function $cosh\, x = \frac{1}{2}(e^x + e^{-x})$ which is the geometric representation of a hanging chain or catenary.

With respect to 'precise' descriptions, it is noted that geometry helps us to understand the heart of the matter. In his wonderful book *On Growth and Form*, D'Arcy Wentworth Thompson (1860–1948) wrote: "*The mathematician knows better than we do the value of an approximation. The child's skipping rope is but an approximation of Huyghens' catenary curve – but in the catenary curve – lies the gist of the matter*" [26]. Such consideration should only be made after the identification step has been successfully made.

Quantities of O-Type

The quantities of O-type are our observations of natural shapes and phenomena, qualitatively and quantitatively. It is important for us in three different ways. First, Gielis Transformations were based on observations in nature, more specifically, of flowers of *Hydrangea*, and square or polygonal stems in plants. Later, various other shapes were described by these transformations. From a mathematical point of view our transformations can be used to describe natural shapes to any degree of precision required. Along the road identifications with other mathematical quantities, formulae and equations were found, looking from many angles and viewpoints. This corresponds to finding connections, and investigating "*whether what is sought is possible or impossible, and how far it is practicable and in how many ways*", our Determinatio or διορισμός step,

We may start from individual observations and identifications, or from the more general theory. We will focus on the motives following the advice of Richard Courant: "*It has always been a temptation for mathematicians to present the crystallized product of their thoughts as a deductive general theory and to relegate the individual mathematical phenomenon into the role of an example. The reader who submits to the dogmatic form will be easily indoctrinated. Enlightenment however, must come from an understanding of motives: live mathematical development springs from natural problems which can be easily understood, but whose solutions are difficult and demand new methods of more general significance*" [27].

Natural problems have been the major source of inspiration for mathematicians since the very beginning. Pythagoras, Newton, Gauss, Riemann and Thom, to name a few, were mathematicians and natural philosophers and their mathematical work was always very closely related to the desire to understand natural phenomena. This is no different in our approach. This book will be about the geometrical and mathematical strategies towards a geometrical theory of morphogenesis.

Here I must pause for a moment, since the strategies are somewhat different from the motives. If I look at plants or natural shapes, I see, feel, smell and experience in every aspect, their development and history. A walk in the park or a forest or a tree nursery is a joy for the senses. Development from seeds or small cuttings to mature plants and the actual presence of the plant, or the recurring seasonal developments of leaves, flowers and seeds as the manifestation of its recent past and the current situation, with a magnificent interaction between the plants and their environment, the action of pollinators, organic and inorganic influences. I experience the energy flowing is this intelligent environment. This is the same for any living and non-living being, form and shape, like the ever-changing forms of clouds or the heavenly bodies, their future and past.

Above all my mind fills with a vision of a most beautiful symphony of life in which rhythms and regularities are the base, and creativity and individuality the actuality. Awe and wonder is the beginning—ἀρχή—of all science. Fortunately, a wide variety of natural phenomena, symphonies and their most individual performances may be described by our mathematical transformations. We can study stills from nature (plants, flowers...) or dynamic phenomena at all scales (time and space). What motivates me is beauty and harmony, caused by awe and wonder about the world we live in, in creatures great and small; what is described in this book is only one language to make sense of this world I experience (Fig. 2.1).

Fig. 2.1 Nature's Grand Symphony

Secondly, identifications between M and O quantities must not rely solely on visual analogy. The statement that Gielis transformations can describe natural and abstract shapes with any required degree of precision cannot rely on a single or a few observations, without actual measurements if we want to use it to study natural shapes and phenomena.

While the origin of our transformations is found in modeling of square bamboos (*Chimonobambusa quadrangularis*), in a wider setting, they must be coupled to repeated and consecutive observations and measurements of natural phenomena and shapes, under widely different conditions and in various closely or less closely related species. Two examples are the modeling of tree rings, and leaves of bamboo.

Thirdly, the very nature of observation is crucial. It has happened many times that I showed shapes to fellow scientists, students, artists and laymen, who classified shapes as round or circular, while they were demonstrably not round (the water fern *Marsilea* in Fig. 2.2). It was only after showing what Lamé curves are that they agreed it was not a circle, but rather a supercircle. To be sure, we have to understand that the circle and rigid geometry is engrained in our minds to such an extent that we may easily be bamboozled by our interpretations.

As one example, the snowflake in Fig. 2.2, would be recognized by many as a truncated triangle, and Fourier analysis would give peaks on period three, but retaining their hexagonal symmetry of ice such snowflakes are actually hexagons with alternating long and short sides, in the same way as a rectangle is a square with alternating short and long sides. For the record, this also relates to the algorithms that we use for automated, so-called unbiased image recognition.

Fig. 2.2 A snowflake with short and long sides and the water fern *Marsilea*

Quantities of M-Type for Natural Shapes

If we wish to study natural shapes and phenomena, it is necessary to agree on some basic definitions and assumptions, many of which may be very different from what we learned. In our approach of transformations, the notions of *unit circle* and *rigidity of rulers* play a crucial role. Our Euclidean geometry is based on a Euclidean circle, with all points on the circle at a fixed distance from a fixed center. As humans we are free to roam in almost any direction (isotropy), and hence we can easily deduce that if we walk in any direction we will arrive after the same interval of time ("as the crow flies"), at points that form a circle. This Euclidean circle is based on our own observations of the world in which we are free to move, at least ideally, and, with the straight line, is at the basis of most of our physical theories and the geometrical tools used.

In the introduction to his book *Minkowski Geometry* [28], however, A.C. Thompson writes: *"Space to Euclid and Newton was uniform and "isotropic", the same in all directions. Such a notion flies in the face of daily experience, where the connotation of up and down is different from that of east to west. There are preferred directions. Another good example is the preferred directions that cause crystals to grow as polyhedra and not spherically like soap bubbles. Unit circles and spheres are not the familiar round objects from Euclidean geometry, but are some other convex shape, called the unit ball"*.

The notion of unit circle or unit ball in nature can be very different from our classic Euclidean circle. To mathematicians and many physicists the notion of metrics and unit circles is well known, being the subject of Minkowski geometry, but it is not (or hardly) known in biology. It is much better understood in crystallography where the so-called Wulff shape provides the unit ball for a given crystallographic configuration, taking into account the molecular structure and the directions of growth, like for example the snowflake. The shapes in Figs. 2.2. and 2.3 are unit circles, and circles in their own right, not deformed or non-circles. Such unit balls introduce a natural anisotropy, with preferred directions of crystal growth.

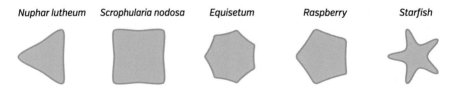

Nuphar lutheum   *Scrophularia nodosa*   *Equisetum*   *Raspberry*   *Starfish*

Fig. 2.3 Some natural shapes as Gielis transformations of a circle

The shapes we will discuss are unit circles (or unit spheres), with the Euclidean circle (or sphere) as one special case. It is possible with our transformations to go from a circle to a square or to a starfish (and back) in a continuous way, and all such shapes will remain unit circles. Likewise we can transform a sphere into a starfish or a pyramid (and back). Since it is continuous, we can describe the smallest possible changes, to discern among many very similar natural objects, like leaves on a tree.

Objective Reality, Mathematics and the Mind's Eye

Henri Poincaré wrote: *"What we call objective reality, in final analysis, is that what is common between several students of this reality, and what could be common to all. This common part cannot be anything else than the harmony expressed by mathematical laws"* [29]. These common parts and the mathematical laws may be different in time, although one must take into account that essentially mathematics, while subdivided in so many fields, has a great unity, especially in relation to the descriptions and observations in nature. When we observe nature and build a mental image to understand what we observe, we are doing science. A simple relation between Ideas I (the mind's eye), Mathematical quantities M (an abstraction of our observations) and Observations O (what we see, feel, here, measure...) is given in Fig. 2.4.

In our understanding we ascend from observing analogies and correspondence (αναλογια or Analogy), to a geometrization (γεωμετρια or Geometry) and finally to understanding the connection between intrinsic and extrinsic characteristics (αρμονια or Harmony), the mathematical laws that govern *"what dwells in the world and the world in which we dwell"* [26].

In the first phase we observe analogies A between widely different phenomena, from which we build a mathematical structure, that may further be enriched with observations. In a subsequent phase a geometrical structure G like Euclidean geometry allowed for many successful laws to describe nature. This mathematical-geometrical structure and quantities allow us to match observations to mathematical

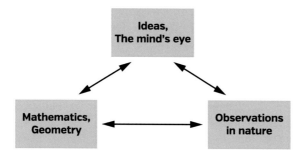

Fig. 2.4 Observing and understanding is a human endeavour

quantities (within experimental errors and statistical boundaries) or allow predictions for certain observations. If observations do not correspond it may be necessary to enlarge or change the desired mathematical structures, whereby in most cases mathematical quantities already exist. Conic sections had been studied long before Kepler and Galilei used them to develop a uniform description of trajectories of stones and planets. Special and general relativity theory built on geometric structures and transformations that had been studied before in mathematics; they could be applied readily so to speak.

There is a third phase as well, the phase of understanding, of harmonizing *H*. This phase occurs when we understand how observations and mathematical laws that describe phenomena well enough, connect whatever we know about the *inner* structure of the phenomena and objects we observe, and their *outer* structure, i.e. the way they relate to the world around them, or how they are embedded in the world around them. Such understanding is found in the relation between intrinsic and extrinsic characteristics of surfaces and manifolds, exemplified in the local by the arithmetic mean and the geometric mean between two principal curvatures in a given point on the surface (The intrinsic nature of a surface or manifold is fixed, but the ultimate shape it will attain is dependent on the environment, the space in which the manifold is embedded), or in the global by the Chern-Bonnett-Gauss theorem which connects the total curvature of surfaces to the half-perimeter of a Euclidean circle or π.

The great theories connect this internal and external, whether in mathematics, physics or biology. The first example is Newtonian theory where curvature of a curve or a surface is correlated to a fixed absolute space (see Schrödinger's quote earlier). The second example from physics is Special Relativity Theory, whereby the physical interpretation by Einstein was the step to harmony (beyond the mathematical use of Lorentz transformations by Lorentz and Poincaré). A fixed Newtonian frame with universal time was replaced by an internal postulate (constancy of speed of light) and an external one (all laws of physics remain valid).

Once this harmony step is achieved, mathematicians and physicists (and biologists) can agree on what they consider (in a certain era) as objective reality, to study everything that we can know or study. D'Arcy Thompson wrote: *"So the living and the dead, things animate and inanimate, we dwellers in the world and the world in which we dwell—πάντα γα μὰν τὰ γιγνωσκόμενα—are bound alike by physical and mathematical law"* [26]. Everything we can know.

Also Alfred Wallace and Charles Darwin's Theory of Natural Selection has the same structure. The biologist Julian Huxley (1887–1975) wrote: *"Darwin was concerned both to establish the fact of evolution, and to discover the mechanism by which it operated, and it was precisely because he attacked both aspects of the problems simultaneously that he was so successful"* [30]. Inherently organisms and species want to develop and multiply ('to their own image, but having the inherent tendency for variation'), but in order to be successful, they have to survive in the environment wherein they live. Neither the inherent nor the external conditions were specified by Darwin, yet his theory proved extremely successful.

A New Pair of Glasses

In this book it will be argued that using semi-rigid rulers to study nature has great value, due to the fact that the semi-rigidity allows us to define a continuous transformation—the Gielis transformations—common to all the phenomena and objects that we will discuss, both in mathematics and in the natural sciences. Our current tools for the study of nature are mainly straight lines and circles, also for studying squares and starfish. We impose "our" geometry onto the study of nature. Our current scientific methods are based on our own human perception of the world. We can build models about nature and its shapes, but computational models able to predict are not enough to understand nature; what is needed are dedicated geometries with intrinsic natural anisotropy through dedicated semi-rigid trans-formations. In particular, it will allow us to connect the isotropic with the aniso-tropic. They will become interchangeable. This requires new pairs of glasses.

Kepler and Newton, Faraday and Maxwell, Poincaré, Lorentz and Einstein, Wallace and Darwin, Mendel and Barbara McClintock [31, 32] offered us new pairs of glasses to study nature, new ways of looking at certain natural phenomena and shapes. Scientific progress is dependent on these glasses as mental instruments. Once more René Thom: "*Les progrès scientifiques sont toujours subordonnées à la possibilité d'un instrument mental qui permette d'exprimer les correspondances, les régularités des choses*" [33]. All scientific progress depends on the availability of a mental instrument that allows for expressing the commonalities, the regularities of things. Description always precedes ideas about the real connection between maths and nature.

Our transformations offer a new, geometric way of looking at nature. They are a unique and uniform way of describing a wide variety of shapes as diverse as plant cells, stems and flowers, starfish, crystals, galaxies and even the relativistic universe itself. Through Gielis transformations curves, surfaces and (sub-) manifolds all become commensurable or symmetrical as conic sections and Euclidean circles, in the spirit of Greek and modern geometry. Besides providing a uniform description, the major change in our point of view is that starfish, crystals and trees will experience the world in totally different ways than humans do, because they are constrained in their way of moving through space and space-time, and hence, if they would have the ability to abstract or geometrize their way of measuring, their geometry of their world would be profoundly different. The unit circle of a starfish could well be a starfish curve, not our circle.

My hope is, apart from contributing towards a geometric (technical) theory of morphogenesis, that this book will further allow for a better understanding of the deep relationships in nature, and that we can use these tools to agree on what objective reality is, beyond anthropomorphism and anthropocentrism. This objec-tive reality can indeed only exist if we accept the independence of observations by organisms and natural objects.

We may hope to get a glimpse of the way other beings, living and non-living, experience, geometrize, symmetrize and harmonize the world in which they live,

Fig. 2.5 Uniting mathematics, science and philosophy

the One World in which we all live (Fig. 2.5). The plant geneticist Barbara McClintock (1902–1992) wrote: *"Basically everything is one. There is no way in which you draw a line between things. What we normally do is to make these subdivisions, but they are not real. Our educational system is full of subdivisions that are artificial, that shouldn't be there"* [32].

Part II
Εκθεσις—Expositio

Chapter 3
−1, −2, −3……, Understand the Legacy

We may picture the product of three thousand years of geometric inventiveness in the form of a tree – the "Geometree" whose roots go back even further and whose branches represent the outcome of centuries of discovery and creation. With or without application, the branches and fruits of this tree are worth contemplating as a remarkable product of human imagination. The Geometree is healthy, vigorous and in full foliage, older than any redwood, and fully as majestic [34].

Robert Osserman

On Arithmetic and Geometric Means

Relations between arithmetic and geometric means will take center stage, so I thought it not amiss to devote a complete chapter to point out the ancient roots of these notions and the relations between seemingly disparate fields.

Exploring the inequality between geometric and arithmetic means is one of the great achievements of the Greek mathematicians. This inequality $GM < AM$ is the corner stone of number theory, for positive numbers. It is related to the problem of transforming a rectangle with sides a and b into a square with equal area (the sides are then given by $GM = \sqrt{ab}$ based on the product) or into a square with the same circumference. In the latter case the side of the square is given by $AM = (a+b)/2$, based on addition. These transformations are also the basis of conic sections and the calculus, as will be shown in this chapter.

When also negative numbers are allowed this inequality can become equality, for example in the product of two functions e^x and e^{-x}. The product of these functions is 1 for any x and so is the geometric mean GM. In only one of the points, however, the geometric mean is equal to the arithmetic mean and that is the lowest point. Indeed, the arithmetic mean of the two functions e^x and e^{-x} is the hyperbolic cosine $cosh(x)$, and equal to one for $x = 0$; One is also the value of the GM for $x = 0$. As $cosh(x)$ is the shape describing a hanging chain or catenary, equality in the GM-AM inequality is achieved precisely in the lowest point of the chain.

© Atlantis Press and the author(s) 2017
J. Gielis, *The Geometrical Beauty of Plants*,
DOI 10.2991/978-94-6239-151-2_3

When the catenary or hanging chain is revolved around the horizontal axis below the chain, a catenoid results. This is the shape a soap film assumes between two rings, when these rings are pulled out of a soap solution. In this case all soap molecules arrange themselves in such a way that not only is the stress evenly distributed over the surface, but the stress is actually zero. This stress is measured in each point on a surface by using the two principal curvatures, which are maximum and minimum curvature (k_1 and k_2), and perpendicular. The product of k_1 and k_2 in this point is the Gaussian curvature, which is the square of the *GM*. In surface theory the *GM-AM* inequality is written as $K \leq H^2$ whereby H is the arithmetic mean of k_1 and k_2. For the catenoid both K and H are zero, as in the plane, the other zero stress surface.

So, in the 18th and 19th century the ideas of Greek mathematicians were gradually transformed into the cornerstone of geometry and physical realizations: minimal surfaces.

The original definition of means by Greek mathematicians is based on ratios and differences for three positive numbers a, b and c. If $a > b > c$, various means were derived by studying the ratio of the differences $\frac{(a-b)}{(b-c)}$, equal to the ratio of the individual numbers. The arithmetic, geometric and harmonic means resulted when the ratio of differences of numbers is equal to $\frac{a}{a} = \frac{b}{b} = \frac{c}{c}$, $a/b = b/c$ and a/c respectively. Our modern notations are different (e.g. *GM* is $\sqrt{ab}$) but the Greek were interested in all combinatorial aspects of numbers and their ratios (Fig. 3.1).

Fig. 3.1 Combinatorial means: *Blue is AM; light Green is GM, Pink is HM, dark green is subcontrary to GM, orange is subcontrary to HM*

ratio	a	b	c
a	a/a	a/b	a/c
b	b/a	b/b	b/c
c	c/a	c/b	c/c

On Derivatives and Nomograms for Means

Geometric and arithmetic means figure in various developments in science, not in the least in the notion of derivatives. In *On Proof and Progress in Mathematics* William Thurston [35] describes how people develop an "understanding" of mathematics and he uses the example of derivatives, that can be approached from very different viewpoints, allowing individuals to develop their own understanding of derivatives.

The derivative can be thought of as:

(1) Infinitesimal: the ratio of the infinitesimal change in the value of a function to the infinitesimal change in a function.

(2) Symbolic: the derivative of $x^n = n \cdot x^{n-1}$, the derivative of $\sin(x)$ is $\cos(x)$, the derivative of f o g is f'o $g * g'$, etc.

(3) Logical: $f'(x) = d$ if and only if for every ε there is a δ, such that when

$$0 < |\Delta x| < \delta, \quad \left| \frac{f(x + \Delta x) - f(x)}{\Delta x} - d \right| < \varepsilon$$

(4) Geometric: the derivative is the slope of a line tangent to the graph of the function, if the graph has a tangent.

(5) Rate: the instantaneous speed of $f(t)$, when t is time.

(6) Approximation: The derivative of a function is the best linear approximation to the function near a point.

(7) Microscopic: The derivative of a function is the limit of what you get by looking at it under a microscope of higher and higher power.

      ......

(37) The derivative of a real-valued function f in a domain D is the Lagrangian section of the cotangent bundle $T*(D)$ that gives the connection form for the unique flat connection on the trivial **R**-bundle D x **R** for which the graph of f is parallel.

Thurston provides this partial list of definitions, to illustrate how 'individual' understanding can differ: *"One person's clear mental image is another person's intimidation. Human understanding does not follow a single path, as a computer with a central processing unit; our brains are much more complex and capable of far more than a single path"* [35]. We should not forget that it has taken mathematicians thousands of years to come to a good understanding of the concept.

Focusing on definition (2) this reduces to a game of cubes and unit elements. Essentially the game component of cubes and beams was very clear to mathematicians from the 16th and 17th century, and many of the other different definitions of derivatives simply follow from these observations. The same procedure underlies important special polynomials. Starting from geometric means, beams, cubes and unit (or neutral) element, the relations between these various fields can easily be shown.

To develop the argument we use a method proposed by Rik Verhulst, a Belgian math teacher whose teaching methods attracted attention from international experts [36–38]. It is a geometrical representation of the nth-arithmetic, nth-geometric and nth-harmonic means. For $n = 2$ the geometric mean *GM* and arithmetic mean *AM* are

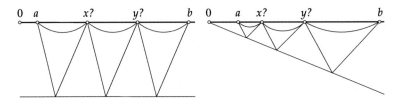

Fig. 3.2 Division of an interval [a, b] into three parts according to *AM* or *GM*

geometry in the sense that they are solutions to one of the oldest optimization problems: For a given rectangle with sides *a* and *b*, *AM* between *a* and *b* is the side of a square with the same *perimeter* as the given rectangle and *GM* between *a* and *b* is the side of a square with the same *area* as the given rectangle with sides *a* and *b*. Coefficient *n* = 2 refers to the use of a square root in *GM*. In general on any straight line, an interval [a, b] can be divided in 2 or more parts following the construction in Fig. 3.2, where the number of intervals is *n* = 3 with the relations given by the Eq. 3.1.

$$
\begin{aligned}
x &= \frac{2a+b}{3} = AM_{\frac{1}{3}} \quad x = \sqrt[3]{a^2b} = GM_{\frac{1}{3}} \\
y &= \frac{a+2b}{3} = AM_{\frac{2}{3}} \quad y = \sqrt[3]{ab^2} = GM_{\frac{2}{3}}
\end{aligned}
\tag{3.1}
$$

The intervals in the left graph show a composition of translations with itself. In the case of *GM* the subsequent divisions define homothetic transformations. An interval can be divided in *n* intervals with *n* = any natural number.

Also the inverse question can be asked: "*Can x and y, z ... be determined graphically using only parallel lines?*" This is straightforward in the case of *AM* but impossible in the case of *GM*. However, a construction using only parallel lines yields the harmonic mean *HM* which is the ratio between *AM* and *GM* (Fig. 3.3).

Fig. 3.3 Graphical construction of the harmonic mean *H* for *n* = 3

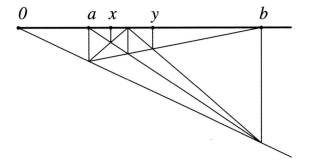

$$x = \frac{3ab}{a+2b} = HM_{\frac{1}{3}} \quad y = \frac{3ab}{2a+b} = HM_{\frac{2}{3}}$$

This leads to the following relations for the division of an interval in three parts (Fig. 3.3).

$$AM_{\frac{1}{3}} \cdot HM_{\frac{2}{3}} = ab, \quad AM_{\frac{2}{3}} \cdot HM_{\frac{1}{3}} = ab, \quad GM_{\frac{2}{3}} \cdot GM_{\frac{1}{3}} = ab \tag{3.2}$$

In general:

$$AM_{\frac{i}{n}} = \frac{(n-i)\,a+ib}{n}, \quad GM_{\frac{i}{n}} = \sqrt[n]{a^{n-i}b^i}, \quad HM_{\frac{i}{n}} = \frac{nab}{ia+(n-i)\,b},$$

$$AM_{\frac{(n-i)}{n}} = \frac{ia+(n-i)\,b}{n}, \quad GM_{\frac{(n-i)}{n}} = \sqrt[n]{a^i b^{n-i}}, \quad HM_{\frac{(n-i)}{n}} = \frac{nab}{(n-i)\,a+ib},$$

$$AM_{\frac{i}{n}} \cdot HM_{\frac{(n-i)}{n}} = ab, \quad AM_{\frac{(n-i)}{n}} \cdot HM_{\frac{i}{n}} = ab, \quad GM_{\frac{i}{n}} \cdot GM_{\frac{(n-i)}{n}} = ab$$

$$\tag{3.3}$$

This method provides for a recursive method for the calculation of roots when using the unit. The nth root $\sqrt[n]{c}$ of a positive number c can be interpreted as $GM_{\frac{1}{n}} = \sqrt[n]{1^{n-i}c^i}$ of the interval $[1; c]$.

Geometric Means and Pascal's Triangle

It is remarkable that these formulae can be generated with simple geometry and algebra, represented in beautiful nomograms, without the sophisticated tools of analysis. It is thus possible to understand various means of different order n geometrically and algebraically [39]. The arguments over which the nth-root is taken are also the various entries of Pascal's Triangle wherein the normal rules of arithmetic are encoded as expansions of $(a+b)^n$. As can be observed, every product between a and b in the Triangle is the argument of the geometric mean of some order between numbers a and b (Fig. 3.4).

Fig. 3.4 First rows of Pascal's triangle

$$
\begin{array}{ccccccccccc}
 & & & & & 1 & & & & & \\
 & & & & 1a & & 1b & & & & \\
 & & & 1a^2 & & 2ab & & 1b^2 & & & \\
 & & 1a^3 & & 3a^2b & & 3ab^2 & & 1b^3 & & \\
 & 1a^4 & & 4a^3b & & 6a^2b^2 & & 4ab^3 & & 1b^4 & \\
1a^5 & & 5a^4b & & 10a^3b^2 & & 10a^2b^3 & & 5ab^4 & & 1b^5 \\
\end{array}
$$

The link with derivatives is straightforward when we observe that in rows the order decreases from n *to* 0, from left to right for a and increases for b in the same direction. If we write $1a^2$ we understand at the same time that this is equal to $b^0 a^2$ and applying the lowering rule to 1, $x^0 \rightarrow 0$. In the one direction we have a lowering of the exponent of a or b, in the other direction an increase according to the following rules:

$$x^n \rightarrow nx^{n-1} \tag{3.4}$$

$$(n+1)x^n \leftarrow nx^{n+1} \tag{3.5}$$

The procedure with Eq. 3.4 is also known as derivation, definition *(2) Symbolic.* Performing this in two directions and using proper normalization dividing by n! the normal binomial coefficients come out.

$$1a^4 \rightarrow 4a^3 \rightarrow 12a^2 \rightarrow 24a^1 \rightarrow 24$$
$$24 \leftarrow 24b \leftarrow 12b^2 \leftarrow 4b^3 \leftarrow 1b^4$$

Multiplying term by term:

$$24a^4 \quad 96a^3 b \quad 144a^2 b^2 \quad 96a^1 b^3 \quad 24b^4$$

Adding all terms and dividing each term by $n! = 4!$

$$1a^4 + 4a^3 b + 6a^2 b^2 + 4ab^3 + b^4$$

So, derivation can be understood directly from simple rules of arithmetic. Many systems in science can be derived from this triangle, starting with the Binomium of Newton, probability theory and combinatorics, the binary system, calculus and the Punnett scheme in genetics.

Tossing a coin has very predictable average outcomes. In this case a and b can be substituted by H(eads)/T(ails). Tossing a coin 3 times gives $2^3 = 8$ possibilities, starting with $H^3 = HHH$ up to $T^3 = TTT$, with all combinations in between. Indeed, we have a total of six 'geometric means' with three times a combination of two H and one T (*HHT, HTH, THH*) and 3 possible combinations of two T and one H (*TTH, THT, HTT*). The outcomes for tossing the coin any number of times n can be read from Pascal's Triangle.

The binary system, the basis of our information technologies, dates back to Leibniz. The first eight numbers of the binary system from 000 to 111 are obtained by $(1+0)^3$, i.e. any combination of three of 1's and 0's, namely 000, 001, 010, 100, 110, 101, 011, 111. Other binary numbers are found by using higher values of the exponent.

The Punnett scheme in genetics for predicting the outcome of crosses with independent genes has the same background. A monohybrid cross with alleles

A and *a* from maternal side and A and *a* from paternal side leads to the combinations AA, aa and 2 times Aa. A dihybrid cross between two double heterozygous parents and unlinked genes gives 16 possibilities in well specified ratios, given by Pascal's Triangle. In genetics, any other outcome would point to for example linked genes. Mendel's laws have a purely arithmetic basis, and for Mendel this felt most natural.

The normal rules of arithmetic have been and still are at the basis of our scientific revolutions. It can even be taken one (or several) step further: Sum the rows of the triangle and make the coefficients general [40]. Summing the first three rows of the triangles and substituting the coefficients of the first three rows by general coefficients A, B, C, D, E and F, gives the general form $Ax^2 + Bxy + Cy^2 + Dx + Ey + F$. When this sum equals zero we obtain the general form for conic sections and the value of the discriminant $(B^2 - 4AC)$ determines whether this defines a circle, ellipse, parabola and hyperbola. It all feels very natural, indeed.

Geometrically: A Game of Cubes and Beams

Geometrically, each entry in a given *n*th row of Pascal's Triangle has the same dimension, and their sum is equal to $(a + b)^n$. The fourth row for example, consists of cubes with side *a* and *b* (and respective volumes a^3 and b^3, one of each), beams with sides *a*, *a* and *b* (and volume a^2b, three of them) and beams with sides *a*, *b* and *b* (and volume ab^2, also three of them).

Any row in the Triangle contains pure n-cubes a^n and b^n (Fig. 3.4, numbers in bold) on the one hand, and *n*-beams on the other hand (Fig. 3.4), whose sum is equal to a hypercube $(a + b)^n$. Hypercubes or *n*-cubes are n-dimensional cubes (with n > 3, with side *a* or *b*), hyperbeams or *n*-beams are n-dimensional beams (with n > 3) of which at least one side is different from all other sides (for example $a^n b^m$) (Fig. 3.5).

What is not known is that it is very easy to turn beams like $a^n b^m$ of dimension $(n + m)$ into cubes of the same dimension. For each hyperbeam $a^n b^m$ one can construct a hypercube with the same volume by taking the $(n + m)$th root of $a^n b^m$, which gives the side of the $(n + m)$-dimensional cube. This is the procedure discussed earlier, on geometric means of a particular order, all of the same dimension. More specifically for $GM_{i/n}$ between two numbers a and b, we specify *i*, *n*, *a* and *b* (either *a* and *b* can be 1)

$$\sqrt[n]{a} = GM_{\frac{1}{n}} \text{ of the interval } [1; a] \text{ since } \sqrt[n]{1^{n-1}a^1}$$
$$= \sqrt{a} \text{ and thus } \sqrt[n+m]{a^n b^m} = GM_{\frac{1}{n+m}} \text{ of the interval } [1; a^n b^m] \tag{3.6}$$

So we can think about these numbers as *geometric numbers*. Simon Stevin (1548–1620) reasoned and thought about geometric numbers (Fig. 3.6). Stevin was

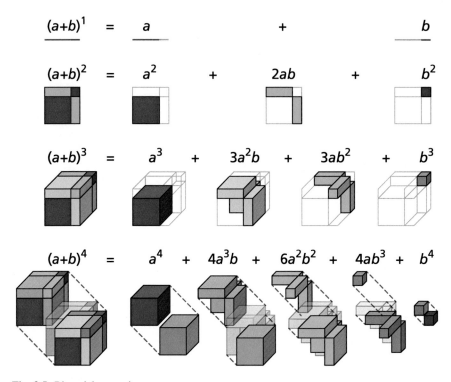

Fig. 3.5 Binomial expansion

one of the greatest mathematicians of the 16th century and his work was both of a pure and applied nature providing a bridge between the old and the new sciences. His equilibrium of forces and the parallelogram rule should be considered as the beginning of abstract algebra and of higher dimensional geometry [41–47].

In Fig. 3.6 examples are given of the powers of 2 (upper row) and the powers of 1 (lower row). 2^3 is a cube, and 2^4 ($= 16$) are two cubes of the size 2^3. Likewise 2^5 ($= 32$) are four of those cubes. For a cube with all sides equal to 1, the results remain the same for any power. It is the neutral element and any number of multiplications of 1 by itself always yields the same result.

An object like b^4 (arithmetically the product $b \cdot b \cdot b \cdot b$) can be understood geometrically in many different ways. The object b^4 is not only a four-dimensional volume of a hypercube with side b, but it is also b times a three-dimensional volume b^3 (side of this cube is b). At the same time it can be seen as $b \cdot b$ times an area of b^2, but b^2 could also be interpreted as a beam of volume $1 \cdot b^2$. This beam can then be made into a cube with the same volume, but with side $(1 \cdot b^2)^{1/3}$, the one-third geometric mean between b^2 and 1, or the second geometrical mean of 1 and b. We have:

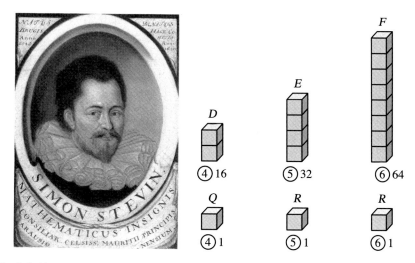

Fig. 3.6 Simon Stevin's Geometric numbers

$$GM_{\frac{2}{3}} \text{ of } [1;b] \text{ is } \sqrt[3]{1^{3-2}b^2} = \sqrt[3]{b^2} \text{ and } GM_{\frac{1}{3}} \text{ of } [1;b^2] \text{ is } \sqrt[3]{1b^2} = \sqrt[3]{b^2} \qquad (3.7)$$

And so on. For the four-dimensional volume, it is important to be reminded how Pascal thought about the fourth dimension: "*Et l'on ne doit pas être blessé par cette quatrième dimension*" [48]. Intelligent people should not be put off by something like the fourth dimension, because in reality it is about multiplication.

A contemporary of Stevin, François Viète (1540–1603) prefers to deal exclusively with numbers avoiding all geometrical connotations. In his "*Logistices speciosae canonica praecepta*" [49] (canonical rules of species calculation), the main law (*Lex homogeneorum*) states that only species of the same kind (*homogeneous* species) can be added or subtracted. In a typical row of Pascal's Triangle, all n-cubes and n-beams are of the same dimension or the same species (*speciosa*), but in general polynomials (for example in one variable $x^4 + x^3 + x$), this is not the case and here the *Lex homogeneorum* rules, according to "common sense", which forbids us to add a hypercube, a cube and a line. This "in fact-not-so-common-sense" is one of the main reasons for the split between arithmetic and geometry.

Contrary to Viète's *Lex homogeneorum* however, it is very easy to get the same dimension for any term of a polynomial using the *unit element*. For example, a polynomial like $x^4 + x^3 + x$ can be written as $(x \cdot x \cdot x \cdot x) + (x \cdot x \cdot x \cdot 1) + (x \cdot 1 \cdot 1 \cdot 1)$; all of the same dimension and actually, all geometric means of different orders between x and the unit element 1 can be written as

$$x^3 + x^2 + x = x^3 + \left(\sqrt[3]{1x^2}\right)^3 + \left(\sqrt[3]{1^2x}\right)^3$$

The Decimal Principle and Fluxions

All this was very natural for mathematicians like Simon Stevin and his contemporaries. The relation between products and rectangles was frequently used for didactical reasons, for example by John Colson in his "perpetual comment" to Newton's *Method of Fluxions and infinite series - To which is subjoin'd: A perpetual comment upon the whole work, consisting of annotations, illustrations and supplements in order to make this treatise A Compleat Institution for the use of learners*" [50].

This treatise brings out a nice historical connection between Stevin and Newton, going back to the complete arithmetical treatment with natural numbers by Stevin in the decimal system in 1585. Stevin showed in his book *De Thiende* [51] that a complete arithmetical control of the real number system is achieved by explicitly demonstrating how all operations on and with real numbers can be carried out when expressing these numbers in the decimal system. As Bartel van der Waerden wrote in his History of Algebra: "*With one stroke, the classical restrictions of "numbers" to integers or to rational fractions was eliminated. Stevin's general notion of a real number was accepted by all later scientists*" [52]. Stevin added in an appendix to *De Thiende* that the decimal principle should be advocated in "*all human accounts and measurements*", thereby anticipating the (partial) realization of this simple idea by two centuries.

Another definition of Stevin in his *L' Arithmétique*, is definition XXVI "*Multinomie algebraique est un nombre consistent de plusieurs diverses quantitez*" [41]. This definition introduces the reader to algebraic multinomials or polynomials. This definition allowed polynomials to be treated as numbers and to be used in normal arithmetic operations. Stevin was the first to realize that code could be treated as data [53].

The importance of the decimal principle and multinomials for geometry and for science cannot be overemphasized. It paved the way for Descartes' algebraic geometry and inspired Newton to write his *Method of Fluxions and infinite series* [54] *with its application to the geometry of curve-line* from the following motivation:

> Since there is a great conformity between the Operations in Species, and the same Operations in common Numbers; nor do they seem to differ, except in the Characters by which they are presented, the first being general and indefinite, and the other definite and particular: I cannot but wonder that no body has thought of accommodating the lately-discover'd Doctrine of Decimal Fractions in like manner to Species, ..., especially since it might have open'd a way to more abstruse Discoveries. But since this Doctrine of Species, has the same relation to Algebra, as the Doctrine of Decimal Numbers has to common Arithmetick: the Operations of Additions, Subtractions, Multiplication, Division and the Extraction of Roots, may easily be learned from thence, if the Learner be but skilled in Decimal Arithmetick, and the Vulgar Algebra, and observes the correspondence that obtains between Decimal Fractions and Algebraick Terms infinitely continued. For as in Numbers, the Places towards the right-hand continually decrease in a Decimal or Subdecuple Proportion; so it is in Species respectively, when the Terms are disposed in an uniform Progression infinitely continued, according to the Order of the Dimensions of any

Numerator or Denominator. And as the convenience of Decimals is this, that all vulgar Fractions and Radicals, being reduced to them, in some measure acquire the nature of Integers, and may be managed as such, so it is a convenience attending infinite Series in species, that all kinds of complicate Terms may be reduced to the Class of simple Quantities...

One of the major developments in the book concerns infinite series. John Colson (1680–1760) who later became Lucasian professor of Mathematics, as one of the successors of Barrow and Newton, wrote in his Introduction: "*As to the Method of Infinite Series, in this the Author opens a new kind of Arithmetick, (new at least at the time of writing this), or rather he vastly improves the old. For he extends the received Notation, making it completely universal and shews, that as our common Arithmetick of Integers received a great improvement by the introduction of decimal Fractions; so the common Algebra or Analyticks, as an universal Arithmetick, will receive a like Improvement by the admission of his Doctrine of Infinite Series, by which the same analogy will be still carry'd on, and farther advanced towards perfection. Then he shews how all complicate Algebraical Expressions may be reduced to such Series, as will continually converge to the true values of those complex quantities or their Roots, and may be therefor be used in their stead*" [50].

Such series also form a positional system. In Taylor series the same rules $Dx^n = n \cdot x^{n-1}$ are key. They give an operational definition of functions with for example, the series for $e^x, \cos x, \sin x$. This is shown in Fig. 3.7 for $\sin x$.

$$e^x = 1 + \frac{x}{1!} + \frac{x^2}{2!} + \frac{x^3}{3!} + \cdots$$

$$\sin x = x - \frac{x^3}{3!} + \frac{x^5}{5!} - \frac{x^7}{7!} \cdots \qquad \cos x = 1 - \frac{x^2}{2!} + \frac{x^4}{4!} - \frac{x^6}{6!} \cdots \qquad (3.8)$$

Fig. 3.7 Sine function for increasing number of terms in partial sums

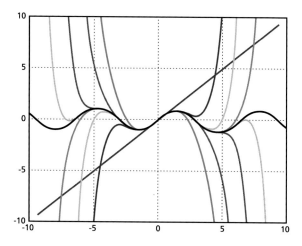

The key to understanding derivatives is putting the unit element back where it belongs [39]. A neutral element is an element that does not influence the outcome (for multiplication this is the unit element), but derivatives demand the substitution of the variable by the unit element. Indeed, performing derivation is substituting x by 1, one at the time in a series. In Eq. 3.6 in the process of derivation (e.g. of x^3) we observe $x^3 = x.x.x \rightarrow 3(x.x.1) = 3x^2$. One cube of x^3 is compared to beams of sides x, x and 1, and we need three of them.

So, if we want to find the derivative of e^x we substitute in every term of the series one factor x by 1 (and if there is no x as in the first term it becomes 0; the "lowering of the exponent operation") and put the original exponent up front lowering $n!$ to $(n − 1)!$. The result is the same series, as expected. It is easy to show that the same occurs if we use the series expansions for sine and cosine, based on the relationship of Euler; $D(\sin x) = \cos x$ and $D(\cos x) = − \sin x$.

In general, derivatives are related to (higher order) *geometric* means between two numbers f and e. Higher order geometric means between two pure numbers f and e involve expressions of the type $(m + n)$th root of $f^m . e^n$ and for $e = 1$: $f^3 + f^2 + f^1 + 1$ can be written as $f.f.f + f.f.e + f.e.e + e.e.e$ (Fig. 3.8). Geometry is the best way to understand derivatives, and the rules of derivation. The derivative of a product $D(x.y)$ can easily be understood by drawing a rectangle with sides $x + dx$ and $y + dy$. Computing the area of that rectangle gives you the exact rule, which according to Vladimir Arnold is impossible to understand from the rules of Leibniz.

$$\text{cub.} = f^3 + 3\, ffe + 3\, fe\,e + e^3$$

Fig. 3.8 From Newton's Principia [55]

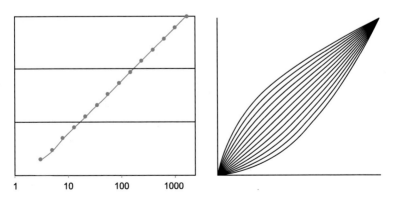

Fig. 3.9 The equal spacing of consecutive Fibonacci numbers in logarithmic scale and the geometry of super- and subparabola in the interval [1; 1] [56]

The Geometry of Parabolas and Allometric Laws

The use of the unit element allows for comparing n-volumes (converting m volumes into n-volumes of the 'same' dimension if needed). This, in our view, also shows that the Greeks were well aware of "units" in relation to conic sections. In a parabola $y = x^2$, the variable y scales to the first power while some other variable x scales to the second, but geometrically a parabola *indicates* that for each coordinate x, one can construct (in Greek terminology to each line with length x it is possible to apply) a square with area equal to x^2, such that this area corresponds exactly to the area of a rectangle with width 1 and height y ($x . x = y . 1$). In this sense the parabola is an *equiareal* figure, a geometric tool for comparing equal areas of different figures [39]. This was known to Ancient Greek geometers and is at the basis of the conic sections.

The parabola and conic sections are based on squares, but we need not stop here. We can raise the variables to any number. Such generalized parabola's were studied in the Renaissance, for example Neil's Parabola with $y = x^{2/3}$ and such higher order parabolas are directly connected to power laws (Fig. 3.9).

Allometric equations or power laws are generally depicted as straight lines in log-log plots, but can be understood in the same geometric way, namely that these equations express some conservation law for n-volumes of n-cubes and n-beams, with the parabola and hyperbola for $n = 2$. The power law $y = x^{3/4}$ or equivalently $y^4 = x^3$ thus states that the 4-cube with side y is exactly the same as the 4-beam with four sides x, x, x and 1 (with volume $x \cdot x \cdot x \cdot 1$). Or $y . y^3 = x^3$. One can build a scale with equal arm length, with on one side y cubes of y^3 and on the other 1 cube of x^3. Or one could choose one cube of each, and then the length of the arms stands in a ratio 1 to y.

An example from physics is Kepler's Law of Periods, which states that the square of the orbital period of a planet is directly proportional to the cube of the semi-major axis of the elliptical orbit. So the volume of a *beam* formed by the orbital period T_1 and the unit element is $[T_1 . T_1 . 1]$ (i.e. the volume of a *beam* with height and length T_1 and width 1), equals the volume of a *cube* $[a_1 . a_1 . a_1]$ (with sides equal to the semi-major axis a_1) up to a constant. On a weighing balance this constant can be interpreted as the length of the arms. This is also an *equiareal* law, like the law of equal areas (Fig. 3.10).

$$T_1^2 = \frac{4\pi^2}{G(M_1 + M_2)} a_1^3 \qquad (3.9)$$

Just as in the solar system, power laws are ubiquitous in living organisms, in so-called allometric equations, expressing the constancy of relative growth [26, 57, 58]. For example, across a broad spectrum of aquatic and terrestrial species (including algae, horsetails, shrubs, trees…) annual growth in plant body biomass per individual G_T is proportional to the ¾ power of the total body mass per individual

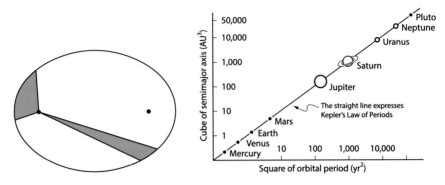

Fig. 3.10 Kepler's law of equal areas (*left*) and law of periods

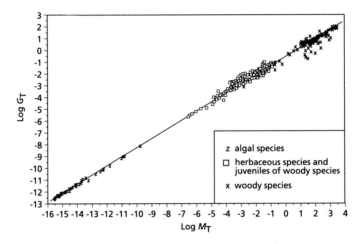

Fig. 3.11 Annual growth G_T is related to total biomass M_T, over a wide range of plant species and growth types. Copyright Karl J. Niklas

M_T (Fig. 3.11) [59]. West, Brown and Enquist derived, starting from energetic balance considerations, a differential equation (the WBE equation) yielding the evolution of biological masses [59]. Any biological evolving complex, follows the logistic paradigm contained in this WBE equation, namely exponential growth and saturation. Allometric laws are ubiquitous in natural systems.

In our days $y = x^{3/4}$ is related often to "fractal dimensions", but these are parabola's of the type $y^n = x^{n-1}$, more specifically $y = x^{3/4} \Leftrightarrow y^4 = x^3$. Or $y \cdot y \cdot y \cdot y = x \cdot x \cdot x \cdot 1$. (Fig. 3.9) Parabola means precise fitting. This in our opinion at least, settles the question whether the Greeks understood derivatives and dynamics in a positive way. It is all about geometric means, which they understood very well.

In a general way the parabolas of the family $y^n = nx^{n-1}$ denote nothing but derivatives. Reading the graphs the other way, $\frac{1}{n}y^n = x^{n-1}$ stands for integration. Simple and pure.

Monomiality Principle for Polynomials

The same simple idea lies at the basis of polynomials and special functions. The explicit form of the series expansion of functions for the exponential, sine and cosine function is:

$$e^x = \sum_{n=0}^{\infty} \frac{1^n}{n!} x^n$$

$$\sin x = \sum_{n=0}^{\infty} \frac{(-1)^n}{(2n+1)!} x^{2n+1} \qquad (3.10)$$

$$\cos x = \sum_{n=0}^{\infty} \frac{(-1)^n}{(2n)!} x^{2n}$$

Sine and cosine are examples of simple polynomials, but in mathematics there are a large number of special polynomials in the theory of special functions, such as Legendre, Chebyshev, Hermite, Jacobi, and Laguerre polynomials. The Chebyshev polynomials (Fig. 3.12) have a very close relationship to circular trigonometric functions and trigonometric identities. Special functions play a crucial role in various fields of mathematics, physics and engineering and during the last decades new families of special functions have been suggested in various branches of

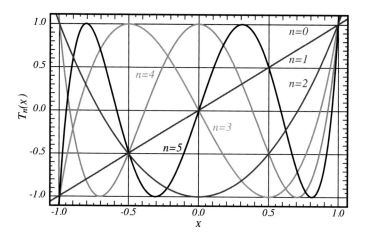

Fig. 3.12 Chebyshev polynomials $T_n(x)$ for $n = 0, \ldots, 5$

physics. In biology Special Functions are rarely used (but this will change, as we will show that certain special functions are rooted firmly in the relation between botany and geometry).

Comparing many such special functions and their explicit form in contemporary pure and applied mathematics to functions of a more elementary nature (e^x, sine, cosine....) from the 17th and 18th century seems somewhat like jumping from older, simpler definitions of the derivative (Thurston's list 1 through 7) to definition 37. Indeed, at first sight contemporary special polynomials are a far cry from the simpler functions and power series, from parabolas and cubes and beams. But is that so?

Most families of special polynomials can be transformed by the same simple procedure of raising and lowering exponents, the same game of means and beams. By virtue of the so-called *monomiality* principle, all families of polynomials, and in particular special polynomials, can be obtained by transforming a basic monomial set by means of suitable operators P and M, called the derivative and multiplication operator of the considered family, respectively. The definition of the *poweroid* introduced by J.F. Steffensen, has been recently framed in the so-called *monomiality* principle by Giuseppe Dattoli, resulting in a very powerful analytical tool for deriving properties of special polynomials [60–63].

The simplest application of the monomiality principle regards the family of monomials $\{x^n\}$ in one variable with derivative and multiplication operators $P = D = d/dx$ and $M = x$ respectively, such that $[P, M] = 1$ and $M(x^n) = x^{n+1}$, $P(x^n) = nx^{n-1}$. These operations simply consist in raising and lowering the exponents of the general monomial belonging to the family. Additional relations are given by $x^0 = 1$ and $Dx^0 = 0$ (One can easily see that it is also possible to start from the antiderivative). The series discussed above are immediate consequences. Using the monomiality principle, the analytical properties of special polynomials can be easily studied. Following this approach, the governing differential equation, recurrence relations and identities can be easily determined. All is based on the very same method of turning rectangles into squares, using basic operations and functions, and all is linked to simple rules of arithmetic and geometric imagination and intuition.

A Way of Thinking...

It is important to understand concepts in different ways. Thurston's list [35] is "*a list of different ways of thinking about or conceiving of the derivative, rather than a list of different logical definitions.*" It is important to be able to look at concepts, for which mathematicians spent thousands of years to come to an ever-better understanding, from very different perspectives, not only tailored to specific talents of individuals, but to safeguard the true spirit of mathematics. Thurston again: "*Unless great efforts are made to maintain the tone and flavor of the original human insights, the differences start to evaporate as soon as the mental concepts are translated into precise, formal and explicit definitions.*" One example is the development of algebra and its deviation from geometric numbers.

The examples of parabola and ellipse show that the Greeks understood the unit element in a general way, not necessarily as a number. Simon Stevin was one of the first to state that the unit was a number "*Que l'Unité est Nombre*" [41]. It becomes the neutral element for multiplication, but it certainly does not need to be "one", as long as the element selected is "neutral". We can understand differentials in the same way. If we look back at Fig. 3.2, and move the point zero as far as possible to the left, and define a unit element close to zero, this element becomes as far as possible removed from a and b, having the least possible influence for computing *AM* or *GM*.

One can call such an infinitely small number for example "*dx*". Obviously, one does not need to move zero; easier is to move the unit element as close as possible to zero. The 'rectangles' xdx in Riemann sums for integrals are a generalization of parabolas, but viewed from a Greek's geometrical perspective, it does not add anything really new. The question whether $dx = 0$ or 1 then becomes pretty irrelevant. It needs to be neutral in multiplication (in which case $dx = 1$), and if it needs to be neutral for addition $(x + dx)$ it has the tendency to be rather close to zero.

From Fig. 3.2 (*GM* and *AM* for $n = 2$) one understands that the geometric mean (*GM*) is strictly smaller than the arithmetic mean (*AM*). In order for *AM* to approach *GM*, the point 0 where the lines cross has to be moved as far to the left as possible. Equality is obtained only in the events when lines run parallel (with the crossing points of the lines situated infinitely far). This is an arithmetical interpretation of Euclid's fifth postulate [64], which has always been considered as having a special status amongst the postulates (Fig. 3.13).

The fifth postulate states that "*If a straight line falling on two straight lines makes interior angles on the same side less than two right angles, the two straight lines, if produced indefinitely, meet on that side on which are the angles less than two right angles*". An alternative is Playfair's axiom: "*In the plane there is one and only one line that goes through a given point* A *and is parallel to a given line* a, *where* A *is not located on* a".

From a Greek perspective the fifth postulate ensures that the arithmetic mean *AM* and the geometric mean GM can be constructed. For positive numbers, based on the geometrical construction of lines that necessarily intersect for any number smaller than infinity, *GM* is strictly smaller than $AM (GM < AM)$, which is the cornerstone of our positive number system. It is an illustration of Chern's remark that "*Euclid's Elements are a geometrical treatment of the number system*". In the 19th century this famous inequality was applied to surface theory, with ramifications from minimal surfaces to space-time models.

Fig. 3.13 Parallel or crossing
lines, with direct
correspondence to the
constructions for *AM* and *HM*

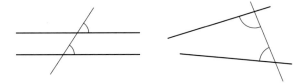

−1, −2, −3......, Understand the Legacy

The formal patterns underlying various means, the normal rules of arithmetic, expansions and special functions are in fact a game of cubes and beams, going back to Greek foundations (−1, −2, −3 is counting down to some time, say −2500 years, when Pythagoreans mastered means, numbers and much more.) With simple principles and elementary functions we can achieve already a very good understanding of mathematics and the relations among various fields. Newton's intuition ("*if the Learner be but skilled in Decimal Arithmetick, and the Vulgar Algebra...*"), is still accurate today, but incomplete. *Basic geometry* is also required for a better and more flexible understanding. To understand higher dimensional cubes and beams as simple geometric numbers, cubes, beams and balances in equilibrium, lifts the ban imposed by François Viète in the 17th century.

It is absolutely forbidden to add the cube x^3 to the square x^2, or to the line x. Up to this very day students are drilled and the ban continues to exist. Even Stephen Hawking maintains on *La Géometrie* of Descartes: "*At the time this was written a^2 was commonly considered to mean the surface of a square whose side is a, and b^3 to mean the volume of a cube whose side is b^3 while b^4, b^5.... were unintelligible as geometric forms.*"

Not only are *n*-cubes with side *a* intelligible but a^n itself carries a lot of valuable geometric information [65]. The coefficients of the binomial expansion of $(2x + 1)^n$ do have a nice geometrical meaning. For example, the tesseract, the four-dimensional cube (x^4) is composed of 16 points, 32 lines, 24 squares, 8 cubes and of course 1 tesseract. Again, this is a nice example of lowering of exponents with a clear geometrical meaning: $(2x + 1)^4 = 16x^4 + 32x^3 + 24x^2 + 8x + 1$. The expansions of 0's and 1's can be used as coordinates of the 16 vertices, with one vertex centered at 0000.

René Descartes (1596–1650) certainly knew about the works of Simon Stevin (1548–1620), not only because they were available in French, but also because of his collaboration with Isaac Beeckman, himself a student of Simon Stevin. The mathematical trinity of algebra, geometry and analysis was understood very well in ancient Greece, where geometry forms the core and the soul. Greek and Hellenistic mathematics and science were advanced in every sense of the word. The rebirth of Eudoxus mathematical findings into Dedekind's cut for defining real numbers is one example. *Euclid's Elements are a geometrical treatment of the number system.* The influence of Stevin's Decimal fractions on the development of fluxions has been pointed out, but this was really based on this very same idea of commensurability. As D.J. Struik (editor of the mathematical parts of the Principal Works of Simon Stevin) writes: "*In his arithmetical and geometrical studies, Stevin pointed out that the analogy between numbers and line-segments was closer than was generally recognized. He showed that the principal arithmetical operations, as well as the theory of proportions and the rule of three, had their counterparts in geometry. Incommensurability existed between line-segments as well as numbers........; incommensurability was a relative property, and there was no sense in calling numbers "irrational", "irregular", or any other name, which connoted inferiority. He went so far as to say, in his Traicté des incommensurables grandeurs, that the geometrical theory of incommensurables, in Euclid's Tenth Book had originally been discovered in terms of numbers, and translated the content of this book into the language of numbers. He compared the still incompletely understood arithmetical continuum to the geometrical continuum already explained by the Greeks, and thus prepared the way for that correspondence of numbers and points on the line that made its entry with Descartes' coordinate geometry*" [41].

Stevin is highly appreciated in the well-educated part of the scientific community, but despite his manifold original contributions Stevin is less well known than other scientists of this period. George Sarton, a great historian of science wrote: "*His greatness is conspicuous and not only in a single domain but in many.... How could people truly admire one whom they do not understand, how could they consider great a man whose greatness they have not yet been educated to appreciate?*" [43]

In some sense, all we are doing is adding some commas and punctuations here and there to what the Greeks have done. Also on the historical side much is to be learned from Plato (that all knowledge is remembrance) and Bacon (that all novelty is but oblivion). One example of such oblivion and "novelty" are the series expansions for cosine and sine, as well as their definitions, traditionally associated with names of 16th and 17th century Western mathematicians. Trigonometric functions sine and cosine, as well as their expansions, were already known to the Indian mathematician Madhava of Sangamagramma (c. 1340–1425) in the early 15th century. Pascal's Triangle was not invented by Pascal, but was known long before in China and Persia [66]. Much of the mathematical, scientific and technological legacy of Ancient Hellas has been forgotten as well [67]. Many special polynomials were invented a century ago and then forgotten, until rediscovered by physics [68]. We should never forget that the history of mathematics is very old and its legacy extremely rich, transcending boundaries in space and time.

Chapter 4
Lamé Curves and Surfaces

*The superellipse has the same beautiful unity as circle and
ellipse but is less obvious and less plain. The superellipse frees
us from the straightjacket of simpler curves such as lines and
planes.*

Piet Hein

On Landscapes and Maps

Two of the defining characteristics of Greek legacy are rational thinking, and the
notion that mathematics has an intimate relationship to the workings of the world.
According to Pythagoras, Everything is Number (and for Pythagoras geometry and
number were inextractably linked). According to Richard Feynman (1918–1988):
*"To those who do not know mathematics it is difficult to get across a real feeling as
to the beauty, the deepest beauty, of nature ... If you want to learn about nature, to
appreciate nature, it is necessary to understand the language that she speaks in.
She offers her information only in one form"* [69]. Fortunately, simple rules are a
basic feature of this language, by whatever name they are known, addition and
multiplication, means, cubes or the monomiality principle.

Despite the simple basic rules contemporary mathematics has evolved into very
diverse landscapes, in pure and applied mathematics or in theoretical physics. Each
uses one or more very specific high-level (increasingly abstract) languages, with
various local dialects. A mathematical concept like curvature goes under a variety
of different names in mathematics and physics, depending on the field. There is the
imminent danger of high-level languages for different landscapes (the languages as
the maps or *des Cartes*), taking precedence over these landscapes and territories
themselves.

When it can be shown that simple rules underlie different mathematical landscapes, the general language (the art of mapmaking) need not be too abstract for a
basic understanding or the development of a certain deeper feeling for the matter.
Geometry rises to glory here. The "maps" we spoke of are based on the
Pythagorean Theorem or triangles in general (including triangulation of surfaces),

© Atlantis Press and the author(s) 2017
J. Gielis, *The Geometrical Beauty of Plants*,
DOI 10.2991/978-94-6239-151-2_4

and geometric means between numbers. The application of the normal rules of arithmetic leads to Pascal's triangle and beyond. One 'beyond' example are conic sections resulting from summing the first rows of the series.

In our Pythagorean world, based on the above rules of arithmetic, there is a clear difference between circles and square. Or polygons. We recall that for modeling natural shapes and phenomena, the geometric means are given preference, and the pure numbers a^n, b^n or, more generally x^n, y^n become "negligible".

In any row of Pascal's Triangle, the likelihood of encountering pure numbers—the left a^n and the rightmost b^n entries of any row—is not high, and they have little influence on the whole of the phenomena. If you throw a coin one hundred times there is always, however small, a possibility that this will result in one hundred consecutive heads, or one hundred consecutive tails. However, such events are very unlikely to happen, and for higher values of n even extremely unlikely. Pure numbers can de facto be neglected, or serve as a unit, having as little influence as possible.

The genius of Pythagoras was that he did not use geometric means to define the circle on the basis of the rectangular triangle, having a "pure number" equation $a^2 + b^2 = c^2$ for the rectangular triangle, or $x^2 + y^2 = R^2$ for a circle, neglecting any geometric means of the type xy or ab. We could now do exactly the same for all rows of the Triangle. Instead of neglecting pure numbers a^n, b^n, they will form the basis of a different kind of geometry or calculus. If we think in this opposite direction, we use only the pure numbers and neglect all geometric means. Take any row in the Triangle, single out the outermost values, the pure numbers and neglect all the rest. The binomial formula $(x+y)^n = \sum_{k=0}^{n} \binom{n}{k} x^k y^{n-k}$ is then rewritten as

$x^n + y^n + [x^{n-1}y + \cdots + xy^{n-1}]$, here without coefficients $\binom{n}{k}$. The part between brackets is now neglected, and we get a kind of modulo form: $x^n + y^n$ modulo geometric means. $(a+b)^n = a^n + b^n$ modulo n is well known in number theory [70].

Gabriel Lamé effectively introduced the curves $x^n + y^n = R^n$ into geometry [8, 71]. These curves include square, circles, asteroids, and many more. Lamé's motivation was the description of crystals in the framework of (Cartesian) geometry. "*Cependant, pour prouver que l'Analyse de Descartes n'est pas incapable de traiter une des plus belles applications du calcul à la Géometrie, je vais indiquer la marche que l'on pourrait prendre. Si je réussis à donner une analyse simple et facilement applicable, je m'applaudirai d'avoir fait rentrer sous le domaine d'un calcul qui doit être général, un sujet qui semblait le fuir*". With the curves named after him, Gabriel Lamé developed a simple yet very general method, based on the higher order binomials or trinomials. In this way he proved that one of the most beautiful applications of calculus in geometry—crystallography—was possible within the framework of Descartes' geometry.

The original motivation of Gabriel Lamé was to open the analytic geometry of Descartes for the description of crystals, of various shapes and symmetries, with sharp corners. He remarked that even if his calculus would be difficult, nothing could be argued against the general applicability. Quite a foresight: since the 1990s we know that such superelliptical shapes exist everywhere in nature. We find supercircles everywhere in nature now, from square bamboos to tree rings, from leaves to flowers, from crystals to galaxies, as single shapes and as signatures of networks. Many examples will be shown in the next chapters. A further generalization in polar coordinates is the Gielis Formula (extending superellipses to different symmetries under the name superformula), the subject of this book, widening the applicability of Lamé's ideas, and at the same time providing simple computational methods.

Here is an important message: We have now seemingly two very different, antagonistic, opposite strategies for studying natural shapes and phenomena, one using geometric means (the classical approach, giving information about the population and the averages), the other one using only pure numbers (Lamé and Pythagorean style separation of variables, giving information about very individual instances; in fact, any particular instance of any sequence of a's and b's is equally rare). But, *these two views are not mutually exclusive.* (My favorite and) rationally the most efficient way is to state that these two ways are *complementary*, they are *dual*.

Lamé Curves and Surfaces

Circles and ellipses, as well as squares and rectangles (Fig. 4.1) are all included in the family of Lamé curves introduced by Gabriel Lamé in 1818, defined by $x^n + y^n = R^n$. More specifically they are all instances of superellipses; these are the planar curves given by Cartesian equations of the form:

$$\left|\frac{x}{A}\right|^n + \left|\frac{y}{B}\right|^n = 1, \tag{4.1}$$

where n is an integer and A and B are positive real numbers. For A = B, the (super) ellipses become (super)circles. In Fig. 4.2, it is shown how Lamé curves are dependent on the exponents being odd or even. The absolute values in superellipses ensure the symmetry in the four quadrants.

This equation can generate a variety of shapes including circles, supercircles, squares and astroids (for A = B), and superellipses and rectangles. All these shapes differ in 3 parameters at most. "Superellipses" is a general name for these shapes, but more specifically subellipses are inscribed in an ellipse, while superellipses circumscribe the ellipse. Likewise subcircles are inscribed in a circle, and supercircles circumscribe a circle. A and B give the size of the circle (for A = B) or the ellipse (for A ≠ B) and the exponent n then transforms this circle or ellipse into other shapes (Fig. 4.1).

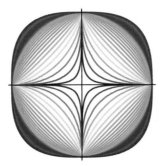
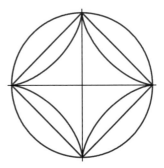

Fig. 4.1 Lamé's supercircles. Copyright Eric Weisstein

Fig. 4.2 Lamé-curves for
$n = 3$ and $n = 4$

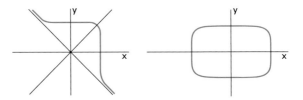

The names sub- and superellipses suggest that the ellipse with exponent $n = 2$ is the central shape, but an inscribed square for $n = 1$ is in fact the shape which divides the area of a circumscribing square in two equal parts and is the exact boundary for exponent n and $1/n$ (Fig. 4.1). Inside the inscribed square, the exponent $n < 1$ (e.g. the red zones in Fig. 4.1), while outside the inscribed square $n > 1$ (e.g. the dark blue zones in Fig. 4.1).

In its most general form, Lamé curves are defined as $x^n + y^n = z^n$. Without absolute values Lamé curves also provide for a unifying description of conic sections [72]:

$$\frac{x}{A} + \frac{y}{B} = 1, \quad \frac{A}{x} + \frac{B}{y} = 1, \quad \left(\frac{x}{A}\right)^2 + \left(\frac{y}{B}\right)^2 = 1, \quad \sqrt{\frac{x}{A}} + \sqrt{\frac{y}{B}} = 1 \tag{4.2}$$

straight line, hyperbola, ellipse and parabola

For the description of natural forms (Fig. 4.3) closed shapes are necessary, hence the use of absolute values in Eq. 4.1.

In June 1853 M. Euzet published a lithographed page with the equation $\left|\frac{x}{A}\right|^m + \left|\frac{y}{B}\right|^n = 1$, with a, b, n and m integers or fractional numbers, positive or negative. This led to an article in *Nouvelles Annales de Mathématiques* with the title *Sur les courbes planes à equations trinômes et les surfaces à equations tétranômes* [73]. The notion *trinômes* and *tétranômes* refer to the number of terms in the equation, three for curves $\left|\frac{x}{A}\right|^m + \left|\frac{y}{B}\right|^n - 1 = 0$. The article itself was written by the

Fig. 4.3 Superelliptical stems of *Phlomis*, *Verbena*, teak and *Euphorbia*

editors of the *Nouvelles Annales* and published one year later (1re Série, vol. 13, 193–200). In this article Lamé's equation is given as example for $n = m$ and in the endnote they refer to Lamé's work.

Traditionally the exponents in Lamé-Euzet curves and surfaces are integers or rational numbers, in line with algebraic curves and equations. Using irrational or real numbers the curves can be considered as *interscendent* curves, a name used by Leibniz for the transition between algebraic and transcendental curves [72]. One example is:

$$\left(\frac{x}{A}\right)^{\sqrt{2}} + \left(\frac{y}{B}\right)^{\sqrt{2}} = 1 \tag{4.3}$$

From Barrow to Lamé

Gabriel Lamé (1795–1870) started studying at the Ecole Polytechnique in 1817, a famous school with teachers like Gaspard Monge, Laplace and Lagrange. In 1818 he wrote a small but influential booklet on methods in geometry: *Examen de différentes méthodes employées pour résoudre les problèmes de géométrie* [8]. In the final part he introduced the equations of plane and lines for the analytic geometrical description of crystals, giving a geometrical foundation to the work of Abbé Haüy, the founder of crystallography. It is then, in the final pages of his book, that Lamé generalizes this to $\left(\frac{x}{A}\right)^m + \left(\frac{y}{B}\right)^m = 1$, for m integer.

With the definition of Lamé curves, Gabriel Lamé gave a very elegant representation of curves that were no longer different individual curves. Henceforth, circle and square, ellipse and rectangle, sphere and cube or cylinder differed in one or a few parameters at most. A whole family of curves and surfaces was defined by one equation, that was a 'multidimensional' version of the Pythagorean Theorem, defining n-cubes and n-beams.

Lamé's work obviously built on earlier researchers. In the 17th century Pierre de Fermat (1601–1665) and Gregory Saint-Vincent (1584–1667) generalized parabola's $y = x^n$ and hyperbola's $y = x^{\frac{1}{n}}$ to determine areas under curves. Power laws defined by power functions $y = x^{\frac{m}{n}}$ are ubiquitous in all fields of science and technology and are geometrically described by superparabola's. Fermat also studied various related equations, esp. the one known from the Last Theorem of Fermat, which states that for $x^n + y^n = z^n$ it is impossible for x, y, z and n to be all >2 and all integer. Felix Klein (1932): "*the assertion of Fermat means now that these curves, unlike the circle, thread through a dense set of the rational points without passing through a single one, except those just noted*" [74]. It would take almost two centuries until Gabriel Lamé gave the complete and precise geometrical interpretation of these curves. The conjecture finally became a theorem when Andrew Wiles gave the final proof in the last decade of the 20th century. Lamé himself noticed the correspondence with his curves and eventually succeeded in proving the Last Theorem of Fermat for $n = 7$.

While Fermat focused on the arithmetic properties, Isaac Barrow studied these curves from a geometrical perspective, one generation later. Barrow's *Geometrical Lectures, Explaining the Generation, Nature and Properties of Curve Lines,* were originally published by the author in Latin, but translated (and revised, corrected and amended) by Isaac Newton. In these lectures Isaac Barrow (1630–1676) demonstrates his tangent method on quasicircular curves (*Cycloformibus*) in the second example given in Lecture X [75].

Barrow gave a geometric interpretation of these curves, but these equations are well known in mathematics, as a particular case of Diophantine equations, as can be found in geometry textbooks of these periods. For example, in *Traité élémentaire de géometrie à deux et trois dimensions* of August Comte in 1843 [76], various examples are given of trinomic equations of the type $y^m + ax^n = b$, with $x^3 + y^3 = 1$ (Fig. 4.2) and more generally $x^{2m+1} + y^{2m+1} = 1$. The development of analytic geometry clarified the relation between these equations and the curves they represent. For example, the equation for the Folium of Descartes is $x^3 + y^3 = 3axy$.

The Last Theorem of Fermat was eventually proven, but there are many problems in Diophantine equations that have similar expressions. In the opinion of Carl Friedrich Gauss: "*Fermat's Theorem as an isolated proposition has very little interest for me, because I could easily lay down a multitude of such propositions, which one could neither prove nor dispose of*" [77]. Indeed, Gauss had the good foresight, since it has become known, in the 1970s that most Diophantine problems are unsolvable. What is interesting here is that Lamé focused on a subset, with the integer exponents and without geometric means. He emphasized the geometrical meaning and use of such curves in the natural sciences. In the future such geometric approach, more restricted than the general case, will help in solving problems and challenges in other fields as well.

The Immortal Works of Gabriel Lamé

Isaac Barrow and Gabriel Lamé are separated by one and a half century. Progress takes time. Introducing shapes (or manifolds) with corners or discontinuities, in a field where deviation from straightness is measured by comparison with circles, has a long history (and the current book has the same goal). Gabriel Lamé was very proud to be able to trace out the path of integrating crystallography, one of the most beautiful applications of mathematics, into analytic geometry. He thus closed the circle for considering two variables raised to arbitrary power, that can either be multiplied (power laws, superparabola and superhyperbola, related to geometric means) or added/subtracted (supercircles, related to arithmetic means).

Gabriel Lamé undoubtedly was one of the leading mathematicians in the first half of the 19th century. For Carl-Friedrich Gauss (1777–1855) Gabriel Lamé was the best French mathematician of this time [78] (and there was quite some competition indeed). When Lamé died in 1860 Joseph Bertrand, secretary of the Académie Française with illustrious members like Christiaan Huyghens, Laplace, Lagrange, Monge, Poisson, and Fourier (and many famous mathematicians ever since) wrote: *"Lamé was a great geometer and he has created methods that are classic today. Above all he was a great mind, a thinker on the most difficult subjects, a dedicated researcher in search of nature's most hidden secrets. No role is greater in the history of science than that of geometrical physicists. This great school has known many illustrious examples within our own Academy, ever since Huyghens. The one we loose today was, in the opinion of all, their most eminent successor"* [79].

Lamé's interests were very broad, and he was both mathematician and engineer (so, difficult to classify) [80]. He was the first to introduce the equation of a plane in the form $\frac{x}{a} + \frac{y}{b} + \frac{z}{c} = 1$. As Coolidge remarks: *"The marriage of algebra and geometry was not without problems in the first centuries after Fermat and Descartes and in the 19th century several geometers proposed different solutions, with Lamé as one of the first"* [81]. For our study of natural shapes it is noteworthy that Fibonacci series were also known as Lamé-series in the 19th and early 20th century [82]. Lamé used this series in relation to the efficient algorithm he developed for computing the greatest communal divisor. In *Mathematics of the 19th century*, edited by A.N. Kolmogorov and A.P. Yushkevic, Gabriel Lamé's work in various fields of mathematics is discussed, and his picture is shown in two of the three volumes [83].

Especially Lamé's work on curvilinear coordinates [84] became very influential [85], and some historians of mathematics even consider this as the real start of differential geometry of surfaces. Lamé's mathematical discoveries are closely linked to his research in the theory of elasticity and mathematical physics. He was the first to apply curvilinear coordinates in space using an orthogonal system, giving the length of an element as:

$$ds^2 = H^2 d\rho^2 + H_1^2 d\rho_1^2 + H_2^2 d\rho_2^2 \tag{4.4}$$

Another great geometer Gaston Darboux (1842–1917) spoke of the immortal works on curvilinear coordinates, *"les immortels travaux de Lamé sur les coordonnées curvilignes"* [86]. Lamé is considered as a father of mathematical physics [87] with the introduction of the differential parameters (invariants) of a scalar field, of first and second kind.

$$\Delta_1 F = \sqrt{\left(\frac{\partial F}{\partial x}\right)^2 + \left(\frac{\partial F}{\partial y}\right)^2 + \left(\frac{\partial F}{\partial z}\right)^2}$$

$$\Delta_2 F = \frac{\partial^2 F}{\partial x^2} + \frac{\partial^2 F}{\partial y^2} + \frac{\partial^2 F}{\partial z^2} \tag{4.5}$$

$$\Delta u = \frac{\partial^2 u}{\partial \rho^2} + \frac{1}{\rho}\frac{\partial u}{\partial \varrho} + \frac{1}{\rho^2}\frac{\partial^2 u}{\partial \vartheta^2}$$

These are the Laplacian, expressed in Cartesian and polar coordinates. The curvilinear coordinates and the differential parameters introduced by Lamé inspired the Italian school of differential geometry with Ricci, Levi-Civita and Beltrami. Vincensini writes: *"C'est à l'instar de G. Lamé, qui a fondé la théorie des coordonnées curvilignes de l'espace euclidien à trois dimensions sur la considérations de deux paramètres différentiels, l'un de 1er ordre et l'autre du 2ème, ouvrant ainsi la voie au calcul différentiel absolu de Ricci et Levi-Civita, que Beltrami a été amené à introduire de son côté, dans la géométrie différentielle des surfaces, les paramètres différentiels qui portent son nom"* [88].

Gabriel Lamé should not only be considered as one of the founders of differential geometry, but also of Riemannian geometry in the opinion of Elie Cartan (1869–1951), the leading geometer of the 20th century. He wrote that the coordinates introduced were not as general as those introduced by Riemann and Helmholtz, but it was Riemannian geometry anyway: *"D'autre part Lamé, quelques années, il est vrai, après la Dissertation de Riemann, avait, pour la géométrie euclidienne à trois dimensions, introduit des coordonnées n'ayant pas un aussi grand degré de généralité que celles de Riemann, mais, dans ce cadre un peu plus restreint, il faisait aussi en somme de la géométrie riemannienne"* [89].

Roughly speaking, an n-dimensional manifold is anything in which points can be defined by an n-tuple of numbers. Curves are examples of which each point can be described by two numbers, the coordinates, or three, for curves in space. If the manifold admits a differentiable structure, it is called a differentiable manifold. In such manifolds one can measure distances and angles and the notion of curvature makes sense. Such measurements can be done on the manifold itself.

The tendency outlined earlier, of using geometric means or pure numbers, are also observed in the two different forms of a Generalized Pythagorean

Theorem resulting either from Lamé curves (in differential form) or from the differential geometry of Gauss:

$$dx^n + dy^n = dz^n \quad g_{11}dx^2 + g_{12}dxdy + g_{22}dy^2 \tag{4.6}$$

The generalized Pythagorean Theorem sensu Lamé uses pure numbers only, while the one sensu Gauss in differential geometry uses geometric means as well.

From Lamé to Square Bamboos

Lamé curves can be used to measure distances in the real world. Cityblock L_1, the Euclidean metric L_2, and the max norm L_∞ are well-known examples of *p-th* root metric of Minkowski (L_p metrics). The taxicab metric is a good example, where taxis need to drive in a system of parallel streets in a rectangular array. These metrics define unit balls (or unit circles in the plane), different from our classic Euclidean circle [28]. Obvious examples for unit balls are Lamé curves. It was one of the main topics of the successors of Gauss and Riemann, namely David Hilbert and Hermann Minkowski. Minkowski created the formal tools to study problems about convex regions and bodies, also in relation to extremal problems and opened the way for the study of norms other than the Euclidean norm on finite dimensional spaces. This was generalized to functional spaces by the Hungarian mathematician F. Riesz around 1913 [90]. Whereas Minkowski geometry is more general than Lamé's approach to describe crystals, it would be logical to call this geometry Lamé-Minkowski geometry.

Due to the work of the Danish mathematician and inventor Piet Hein, Lamé curves became very popular in the 1960s, amongst others with articles of Martin Gardner in his series in the *Scientific American*, later published in the book *Mathematical Carnival* [65] and a great article in *Life Magazine* (1966) "Piet Hein bestrides arts and science" [91]. In *The Mathematical Gazette*, N.T. Gridgeman wrote: "*Piet Hein, the contemporary Danish poet-designer-scientist (and inventor of mathematical games) rediscovered the curves and has been using "superellipses" (Lamé curves with n > 2, and therefore oval) in objets d'art, furniture, pottery, fabric patterns, and so on. But his major achievement to date is a sunken oval shopping plaza, promenade, and pool in the center of Stockholm...*" [92].

For Piet Hein the superellipse was an iconic solution between a round and a square worldview: "*Man is the animal that draws lines, which he himself then stumbles over. In the whole pattern of civilization there have been two tendencies, one toward straight lines and rectangular patterns and one toward circular lines. There are reasons, mechanical and psychological, for both tendencies. Things made with straight lines fit well together and save space. And we can move easily – physically or mentally – around things made with round lines. But we are in a straitjacket, having to accept one or the other, when often some intermediate form would be better. To draw something freehand – such as the patchwork traffic circle*

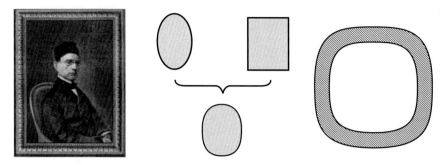

Fig. 4.4 Gabriel Lamé, superellipses combine the advantages of rectangle and ellipse, and cross section of supershaped stem or tube

they tried in Stockholm – will not do. It isn't fixed, it isn't definite like a circle or square. You don't know what it is. It is not esthetically satisfying. The superellipse solved the problem. It is neither round nor rectangular, but in between. Yet it is fixed, it is definite – it has a unity. The superellipse has the same beautiful unity as circle and ellipse but is less obvious and less plain. The superellipse frees us from the straightjacket of simple curves such as lines and planes" [91].

Piet Hein was also famous for his *Grooks,* small, illustrated poems. One of my favorite ones is the grook that warns against megalomania: *"The Universe may be as big as they say, but it would not be missed, if it didn't exist".*

The superelliptic shape of Sergel's Torg combines the power of the rectangle to optimize the area of the shopping center and the power of the ellipse to optimize the traffic flow (Fig. 4.4). The same principles were observed in many plants, starting with square bamboos [93]. Certain species of bamboos of the genus *Chimonobambusa*, have square-like stems along the whole length of the culms, or only in the lower part. They grow in temperate zones, at heights between 500 and 1800 m of altitude, with annual mean temperatures of 8–20 °C, with minimum temperatures up to −15 °C. One species, *Chimonobambusa quadrangularis* has particularly superelliptical stems. The name *Chimonobambusa* refers to the fact that the culms develop late in the season (Chimono is autumn) (Fig. 4.5).

The representation which led to Gielis curves originates in the study of flowers, where the shape of petals or sepals can be, like in the leaflets of *Marsilea* (Fig. 2.2) arranged in a "square", rather than a circle. This fact, unnoticed by botanists, was originally found in Hydrangea (Fig. 4.6).

Many plants have square stems, most notably Lamiaceae and Melastomataceae [94]. This is related to decussate phyllotaxy, but also to optimization principles, related to space use or resistance against bending or torsion. One notable example is teak (*Tectona grandis*) in which all young stems and branches are square, supporting large leaves. Square stems are also found in various succulent Euphorbias and Asclepiads. Lamé ovoids are found in stem succulents and pachycaul trees, like *Jatropha berlandieri*, *Calibanus hookeri*, *Pachypodium brevicaule* and *Nolina*. One finds many examples in nature with fourfold symmetry, in bacteria, diatoms, plant anatomy, the shape of Bénard cells in hydrodynamics, up to the shape of galaxies.

Fig. 4.5 *Chimonobambusa quadrangularis, Dendrocalamus giganteus (culm),* and *Silphium perfoliatum* (flower and stem)

Fig. 4.6 Hydrangea sepals arranged in a square

Lamé Surfaces and Superquadrics

The generalization of Lamé-Euzet curves to 3 dimensions or more is straightforward. For three dimensions we have Lamé-Euzet surfaces:

$$\left(\frac{x}{A}\right)^m + \left(\frac{y}{B}\right)^n + \left(\frac{z}{C}\right)^p = 1 \tag{4.7}$$

In the late 20th century these surfaces were introduced in the field of computer graphics, as superquadrics [95, 96]. With slight variations, supertoroids and superhyperboloids of one or two sheets could be defined. The cross sections of toroids or hyperboloids could be made in any shape defined by Lamé curves.

Gabriel Lamé himself went from squares to hexagons and into 3D by combining shapes to describe crystal shapes. One example of combinations of 3D-superspheres is due to Onaka, originating in material sciences to describe shapes of nanoparticles of metals [97, 98]. He defined polyhedral superspheres $[h(x,y,z)]^{\frac{1}{p}} = R$ with $p \geq 2$ (Eqs. 4.8) as follows (whereby hexa, octa and dodeca refer to the number of planes):

1. Cube or Regular hexahedral supersphere

$$h_{hexa}(x,y,z) = |x|^p + |y|^p + |z|^p$$

2. Octahedron or Octahedral supersphere

$$h_{octa}(x,y,z) = |x+y+z|^p + |-x+y+z|^p + |x-y+z|^p + |x+y-z|^p$$

3. Rhombic − dodecahedral supersphere

$$h_{dodeca}(x,y,z) = |x+y|^p + |x-y|^p + |y+z|^p + |y-z|^p + |x+z|^p + |x-z|^p$$

4. Polyhedral supersphere

$$\left[h_{hexa}(x,y,z) + \frac{1}{a^p}h_{octa}(x,y,z) + \frac{1}{b^p}h_{dodeca}(x,y,z)\right]^{1/p} = R$$

$$(4.8)$$

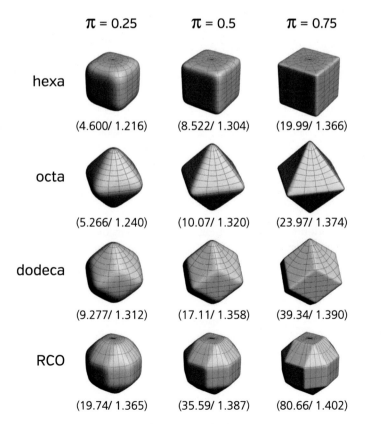

	π = 0.25	π = 0.5	π = 0.75
hexa	(4.600/ 1.216)	(8.522/ 1.304)	(19.99/ 1.366)
octa	(5.266/ 1.240)	(10.07/ 1.320)	(23.97/ 1.374)
dodeca	(9.277/ 1.312)	(17.11/ 1.358)	(39.34/ 1.390)
RCO	(19.74/ 1.365)	(35.59/ 1.387)	(80.66/ 1.402)

Fig. 4.7 Polyhedral superspheres. Copyright Susumu Onaka

The choice of parameters a and b provides a weight factor for adding the octahedron and dodecahedron to the cube. In Fig. 4.7 the influence of a and b is visualized. The shapes are given for $p \rightarrow \infty$. It is noted that in general algebraic expressions for 2D or 3D shapes are not simple. Simple sums however, are not entirely satisfactory since in combining shapes, the zones where the original shapes meet (two or more) may be transition zones with sharp edges and cusps. By moderating the exponent p the edges can be rounded and this example in nano-materials is only one of many applications of Lamé's ideas of superellipses and superquadrics. Lamé's fingerprints are everywhere in nature [99].

Chapter 5
Gielis Curves, Surfaces and Transformations

> *Describing form is one of the more intractable problems in biology. Researchers have come up with many ways to describe leaves and shells, for example, but there is little unity: Things have become cumbersome and idiosyncratic. The Superformula might provide a single, simple framework for analysing and comparing the shapes of life. This is an exciting development.*

> Karl J. Niklas

Lamé Curves in Polar Coordinates

In *Lamé Ovals,* Gridgeman writes: *"The set of Lamé curves for A = B and n > 1 fills the area between a square and its inscribed circle. A further generalization that immediately suggests itself is the coverage of the analogous curves that lie between other regular polygons and their inscribed circles. This is merely a remark"* [92]. Indeed, one major drawback of Lamé curves, supercircles and superquadrics is their limitation to squares and fourfold symmetry, describing circles, squares and rectangles, cubes and spheres, but not triangles, pentagon or octahedrons. This challenge can be addressed when rewriting Lamé curves in transcendental functions, via polar coordinates. The step to the Superformula or Gielis curves and transformations is a logical extension of Lamé curves, in which the concept of supercircles is generalized to include symmetries for any real number, integer or non-integer.

Lamé superellipses are given in *polar co-ordinates* ρ and ϑ by using the substitution $x = \rho(\vartheta) \cos \vartheta$; $y = \rho(\vartheta) \sin \vartheta$ [1, 100]:

© Atlantis Press and the author(s) 2017
J. Gielis, *The Geometrical Beauty of Plants*,
DOI 10.2991/978-94-6239-151-2_5

$$|x|^n + |y|^n = R^n$$

$$\downarrow$$

$$|\rho(\vartheta)\cos\vartheta|^n + |\rho(\vartheta)\sin\vartheta|^n = R^n$$

$$\downarrow$$

$$\rho^n(\vartheta)[|\cos\vartheta|^n + |\sin\vartheta|^n] = R^n \tag{5.1}$$

$$\downarrow$$

$$\rho^n(\vartheta) = \frac{R^n}{[|\cos\vartheta|^n + |\sin\vartheta|^n]}$$

$$\downarrow$$

$$\rho(\vartheta) = \frac{R}{\sqrt[n]{|\cos\vartheta|^n + |\sin\vartheta|^n}}$$

The circle is a special case for $n = 2$, with equation $\rho(\vartheta) = R$ but supercircles have the same "structure" as the Pythagorean Theorem. Actually $\rho(\vartheta)$ becomes a "flexible" radius, a ratio between the radius of the basic circle R and the unit supercircle (5.1).

Starting from ellipses we obtain (Fig. 5.1):

$$\varrho(\vartheta) = \frac{1}{\sqrt[n]{\frac{1}{A}|\cos\vartheta|^n + \frac{1}{B}|\sin\vartheta|^n}} \tag{5.2}$$

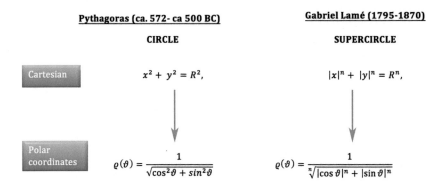

Pythagoras (ca. 572- ca 500 BC)	Gabriel Lamé (1795-1870)
CIRCLE	SUPERCIRCLE

	CIRCLE	SUPERCIRCLE
Cartesian	$x^2 + y^2 = R^2,$	$|x|^n + |y|^n = R^n,$
Polar coordinates	$\varrho(\vartheta) = \dfrac{1}{\sqrt{\cos^2\vartheta + \sin^2\vartheta}}$	$\varrho(\vartheta) = \dfrac{1}{\sqrt[n]{|\cos\vartheta|^n + |\sin\vartheta|^n}}$

Fig. 5.1 From Pythagoras and the circle, to Lamé's supercircles

Generalized Pythagorean Theorems

In the same way as cosines and sines are defined on the circle, Lamé sines and cosines ($_n$sin and $_n$cos) are defined on supercircles. This leads in a natural way to a generalization of the Theorem of Pythagoras for Lamé curves [101, 102], with the classical Theorem for $n = 2$

$$(_n \cos \vartheta)^n + (_n \sin \vartheta)^n = 1, \tag{5.3}$$

with:

$$\varrho(\vartheta) = {}_n\cos \vartheta = \frac{1}{\sqrt[n]{|\cos \vartheta|^n + |\sin \vartheta|^n}} \cdot \cos \vartheta,$$
$$\varrho(\vartheta) = {}_n\sin \vartheta = \frac{1}{\sqrt[n]{|\cos \vartheta|^n + |\sin \vartheta|^n}} \cdot \sin \vartheta \tag{5.4}$$

The stretched radii and the coordinate functions are very shape specific. One invariant is the halflength of the curve, a specific number we call π_n in case of Lamé curves [102]. For $n = 2$, the Lamé curve is the circle and $\pi_{n=2} = \pi$. An infinity of beautiful equations, relations, and geometries can be derived from this simple principle, but they are all interconnected.

The Folium of Descartes can be considered as an historically early version of 5.1, as a transformation acting on a function $f(\vartheta)$, rather than on a constant function R. In this case the function is $f(\vartheta) = 3a \sin \vartheta \cos \vartheta$, for $n_{2,3} = 3$ and $n_1 = 1$.

$$\rho(\vartheta) = \frac{1}{\cos(\vartheta)^3 + \sin(\vartheta)^3} (3a \sin \vartheta \cos \vartheta) \tag{5.5}$$

In Cartesian coordinates the Folium is defined as $x^3 + y^3 = 3axy$, from which $\frac{x^3 + y^3}{3axy} = 1$. If we convert to polar coordinates by directly substituting $x = \cos \vartheta$ and $y = \sin \vartheta$ we obtain the equation of the Folium in polar coordinates [72].

Equation 5.5 can also be read as the product of two coordinate functions of a Lamé curve as follows:

$$3a \frac{\cos \vartheta}{\sqrt{\cos(\vartheta)^3 + \sin(\vartheta)^3}} \cdot \frac{\sin \vartheta}{\sqrt{\cos(\vartheta)^3 + \sin(\vartheta)^3}} \tag{5.6}$$

Here we have the same idea: what is 1 in Cartesian coordinates can be considered as a "stretchable" unit element $\rho(\vartheta)$, dependent on ϑ. The Folium is one of the rare instances where Cartesian and polar expressions are "one-on-one". In general this is not straightforward at all. We will develop a method later for a one-on-one conversion for supershapes from polar to Cartesian and back.

From Lamé to Gielis Curves

The exponent n is applied to the trigonometric functions individually, and to the root over the sum of both. It is easy to envisage, like Euzet did, that these can have different values, increasing the degrees of freedom. We denote the exponents over the cosine and sine term by n_2 and n_3 respectively, and n_1 over the sum of both as $\frac{1}{n_1}$. For $n_2 = n_3$ and $n_{2,3} < 1$, the shape is inscribed in the base circle $n_{2,3} = 2$. For $n_{2,3} > 2$ the shape will circumscribe the circle. The parameter $\frac{1}{n_1}$ then acts as a pull or push force on the sides of the shape, while the anchor points of the shape to the circle (at $0°$, $90°$, $180°$, $270°$) will remain fixed.

$$\varrho(\vartheta) = \frac{1}{\sqrt[n_1]{(|\cos\vartheta|)^{n_2} + (|\sin\vartheta|)^{n_3}}} \tag{5.7}$$

Introducing a coefficient $m/4$ that allows more particular symmetries of rotation around O than those related to the four quadrants of the Cartesian co-ordinate system, and increasing the degrees of freedom in the exponents occurring in the three terms of Lamé curves in polar coordinates, we obtain [1, 2, 100]:

$$\varrho(\vartheta; m, A, B, n_1, n_2, n_3) = \frac{1}{\sqrt[n_1]{\left|\frac{1}{A}\cos\left(\frac{m}{4}\vartheta\right)\right|^{n_2} + \left|\frac{1}{B}\sin\left(\frac{m}{4}\vartheta\right)\right|^{n_3}}} \tag{5.8}$$

with $A, B, n_1 \in \mathbb{R}_0; m, n_{2,3} \in \mathbb{R}$. Examples are shown in Fig. 5.2.

The parameters can be real numbers, not necessarily integers as in Lamé curves. The values of m can be different on both cosine (m_1) and sine (m_2) terms and the two terms can be added or subtracted. Indeed, a minus sign allows for describing curves with indefinite metrics based on hyperbolas rather than on ellipses. Interestingly, the base of the natural logarithms and the catenary, the illustrious number e, is based on the hyperbola. A unit "invariant under change" is defined as the area under a hyperbola having value 1, running from $+1$ to the number corresponding to e. If we start from $x^2 - y^2 = 1$, and follow the same route as (5.1) we get a minus sign in (5.9). More generally for $A, B, n_1 \in \mathbb{R}_0; m_{1,2}, n_{2,3} \in \mathbb{R}$, we have:

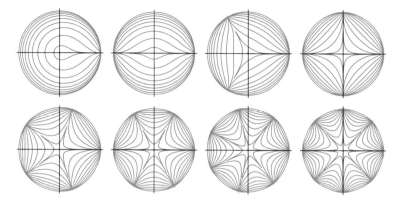

Fig. 5.2 The symmetry parameter m increases from 1 to 8 for $n_{2,3} < 2$ [103]

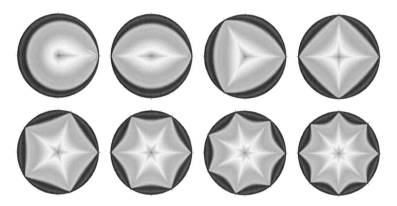

Fig. 5.3 Supershapes in continuous color gradients [104]

$$\varrho(\vartheta; A, B, n_1, n_2, n_3) = \frac{1}{\sqrt[n_1]{\left|\frac{1}{A}\cos\left(\frac{m_1}{4}\vartheta\right)\right|^{n_2} \pm \left|\frac{1}{B}\sin\left(\frac{m_2}{4}\vartheta\right)\right|^{n_3}}} \qquad (5.9)$$

To describe points inside or outside the curves, Eqs. 5.8 and 5.9 can also be written as inequalities, using $\varrho(\vartheta) \leq$ or $\varrho(\vartheta) \geq$ instead of the equality. In this case each point inside and outside the shape is defined, and consequently height, colors, or density gradients can be described as well; a homogeneous blue color or a gradient of different colors, both are possible (Fig. 5.3).

Those curves will further be referred to as *supershapes*, in analogy to superellipses or supercircles. Equations 5.8 and 5.9 were originally called *Superformula*.

After the initial publication (in botany), in mathematics supershapes were renamed as *Gielis curves*. In scientific literature one finds also names like *Gielis Formula GF* or *Gielis Superformula GSF*.

Superpolygons and Regular Polygons

With the symmetry variable m in Eqs. 5.8 and 5.9, zerogons ($m = 0$; no-angles), monogons ($m = 1$; one-angles), and digons ($m = 2$; two-angles), as well as triangles ($m = 3$), squares ($m = 4$) and polygons with higher rotational symmetries can be defined. The parameter m allows the orthogonal axes used in superellipses, to fold in or out like a fan and determines the number of points fixed on the unit circle (or ellipse for $A \neq B$) and their spacing. These points will always remain fixed.

The values of n_2 and n_3 (for $n_2 = n_3$) determine whether the shape is inscribed in or circumscribing the unit circle. For $n_2 = n_3 < 2$ the shape is inscribed (subpolygons), while for $n_2 = n_3 > 2$ the shape will circumscribe the circle (superpolygons). The value of n_1 will further determine the shape and acts like a gravitational pull force away from or towards to the unit circle. Corners can be sharpened or flattened and the sides can be straight, convex or concave (Table 5.1 [102]).

Subpolygons are inscribed in the circle and rotated by π/m relative to superpolygons, circumscribing the circle. Interestingly, when subpolygons transform into superpolygons (and vice versa), vertices transform into edges, and edges into vertices, because of the fixed points on the unit circle. Similar closed shapes are generated which close after one rotation (0–2π) by selecting zero or a positive integer for m. Exactly the same shape is generated for every subsequent rotation by 2π. When further changes of Eq. 5.8 are applied, such as when A and B are not equal, shapes as in Table 5.1, column 8 result.

Lamé curves give a precise expression for squares (inscribed square for $n = 1$ and circumscribed for $\lim_{n\to\infty} (|x^n| + |y^n|)$). To the family of superellipses belong the rhombs (inscribed in the ellipse) and rectangles (circumscribing the ellipse). Equation 5.8 allows for a precise expression for regular polygons for $m > 4$ (Fig. 5.4) [102, 105]:

$$\varrho(\vartheta) = \lim_{n_1 \to \infty} \frac{1}{\left[\left|\cos\left(\frac{m}{4}\vartheta\right)\right|^{n_2} + \left|\sin\left(\frac{m}{4}\vartheta\right)\right|^{n_3}\right]^{1/n_1}} \quad \text{with} \quad n_1 = \frac{n_2 - 2}{2\log_2\left(\frac{1}{\cos\frac{\pi}{m}}\right)} \quad (5.10)$$

or:

$$\varrho(\vartheta) = \lim_{n_1 \to \infty} \frac{1}{\left[\left|\cos\left(\frac{m}{4}\vartheta\right)\right|^{2(1-n_1\log_2\cos\frac{\pi}{m})} + \left|\sin\left(\frac{m}{4}\vartheta\right)\right|^{2(1-n_1\log_2\cos\frac{\pi}{m})}\right]^{1/n_1}} \quad (5.11)$$

Table 5.1 Superpolygons and regular polygons. Copyright Kristof Lenjou

m	$n_1 = n_2$ $= n_3 = 1$	$n_1 = n_2$ $= n_3 = \frac{1}{2}$	$n_1 = n_2$ $= n_3 = \frac{3}{2}$	$n_1 = 5,$ $n_2 = n_3 = 20$	$n_1 = 20,$ $n_2 = n_3 = 5$	$n_1 \neq n_2 \neq n_3$	$n_i = 1,$ $a = 2$
1							
2							
3							
4							
5							
6							
7							
8							
9							
10							

Polygrams and Generic Symmetries

Equations 5.8 and 5.9 also work for non-integer rational m. When m is not an integer, the shape generated does not close after one rotation. If $m = p/q$ is a rational number, the shape will close after a number of rotations equal to q and the number of 'angles' is p. Such shapes may be called "*superpolygrams*", in analogy with "polygrams" or Rational Gielis curves RGC.

For $m = 5/2$ for example, the shape will close with five angles after only two rotations and will then have 5/2 or 2.5 angles in one rotation, spaced 144° or $4\pi/5$ apart (Fig. 5.5). For $m = 8/3$ the shape will close after three rotations and will be repeated every 6π. The rational numbers 5/2 and 8/3 are ratios of Fibonacci numbers F, more specifically the ratio of F_{n+2} and F_n with n indicating the position of the number F_n in the Fibonacci sequence. This is the inverse of the ratios used to describe Fibonacci patterns in plant phyllotaxy F_n/F_{n+2} (such as $\frac{2}{5}; \frac{3}{8}; \frac{13}{5}; \frac{21}{8} \cdots$;

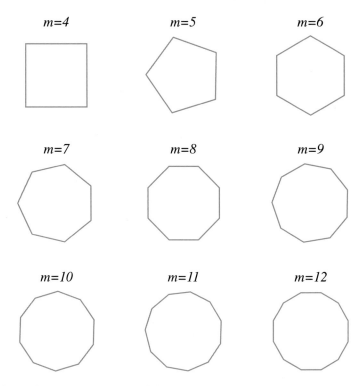

Fig. 5.4 Regular superpolygons. Copyright Diego Caratelli

Fig. 5.5). A Fibonacci polygram will be generated for the $\lim\limits_{n\to\infty}(F_n/F_{n+1}) = \frac{\sqrt{5}+1}{2}$, the golden ratio Φ and the ideal phyllotactic angle 137.5...°. There will be no repeating pattern using irrational numbers, but, for example, a π-gon will have on the average π angles per rotation. We prefer m rather than $1/m$ since our super-shapes generate the genetic spirals.

For a *pentagon* and *superpentagons* this is equal to $k.72°$. For a *pentagram* and *superpentagrams* with generic symmetry 5/2, the angles are spaced 144° apart

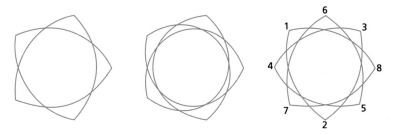

Fig. 5.5 Polygrams with m = 5/2; 5/3 and 8/3

Fig. 5.6 Phyllotactic patterns in plants [1]

(Figs. 5.5 and 5.6). In two rotations the shape closes and the number of angles for pentagon and pentagram is equal. The axes of symmetry for regular polygons pass through the origin and $k.2\pi/m$, and through the origin and $k.\pi/m$ with k integer.

One example of "*generic symmetry*" concerns superpolygons with alternating short and long sides. In this case the classical view would assign different symmetries, as in the case of a rectangle versus a square. In Lamé superellipses however, we have both squares (for R) and rectangles (for $A \neq B$), whereby the edges can be equal or alternating long and short sides. This can be extended with Eq. 5.8. Indeed, it is only in a superficial way that these curves have a different symmetry. In Fig. 2.2 the "hexagonal" molecular symmetry of ice is retained in seemingly "triangular" snowflakes. If $m = 6$ for example, for $A \neq B$ a hexagon with alternating short and long side results starting from an ellipse, and the snowflake's symmetry remains six, corresponding to its molecular symmetry.

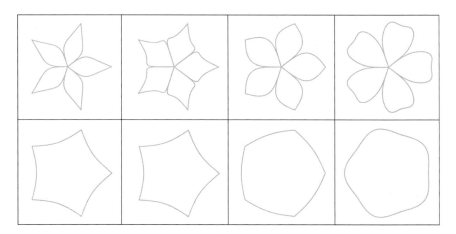

Fig. 5.7 Choripetalous five-petalled flowers with the corresponding constraining superpolygons

Gielis Transformations

Gielis curves can be interpreted starting from *the unit circle centered at* 0 ($\rho = 1$) by the *transformation* for any choice of parameters A; B; m; n_1; n_2; n_3. The superformula *transforms* the unit circle into various sub- and supercircles. The reverse is also true: all supercircles are (unit) circles in their own right. Instead of transforming the unit circle, all planar curves determined by polar equation $\rho = f(\vartheta)$, whereby $f(\vartheta)$ can be any positive real function, can be transformed into planar curves with polar equations [100]:

$$\varrho(\vartheta; f(\vartheta), A, B, n_1, n_2, n_3) = \frac{1}{\sqrt[n_1]{\left|\frac{1}{A}\cos\left(\frac{m}{4}\vartheta\right)\right|^{n_2} \pm \left|\frac{1}{B}\sin\left(\frac{m}{4}\vartheta\right)\right|^{n_3}}} f(\vartheta) \qquad (5.12)$$

Transforming a circle with radius R is obtained for $f(\vartheta) = R$. Two logical functions to transform are spirals and trigonometric functions. Trigonometric functions are closely related to flowers, since Guido Grandi (1671–1742), in a letter to Leibniz in 1713, described rose curves or *Rhodonea* curves as $\rho(\vartheta) = \cos m\vartheta$. These Rhodonea or rose curves actually were the inspiration to include the $m/4$ symmetry parameter.

When our transformation is applied to Rhodonea curves, superflowers or superroses result. For the flowers in Fig. 5.7, the petal number is determined by $\cos m\vartheta$ and their size by n_4. In a classic rose curve the number of petals is m for m odd and $m/2$ for m even. By using absolute values on the numerator, the number of petals is the value of m, for m odd or even.

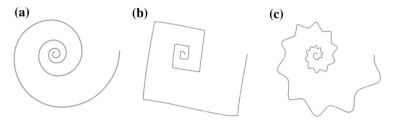

Fig. 5.8 $\rho(\vartheta) = e^{0.2\vartheta}$ with **b** $m = 4$ and $n_1 = n_2 = n_3 = 100$ and **c** $m = 10$ and $n_1 = n_2 = n_3 = 5$

$$\varrho(\vartheta; f(\vartheta), A, B, n_1, n_2, n_3, n_4) = \frac{\left|\cos\frac{m}{2}\theta\right|^{1/n_4}}{\sqrt[n_1]{\left|\frac{1}{A}\cos\left(\frac{m}{2}\vartheta\right)\right|^{n_2} \pm \left|\frac{1}{B}\sin\left(\frac{m}{2}\vartheta\right)\right|^{n_3}}}$$

These superroses (or supercosines and supersines) are the coordinate functions of supershapes. The distance from the origin to the perimeter of the flower $\varrho(\vartheta)$ is determined by the *product* of $|\cos m\theta|^{n_4}$ and Eq. 5.8 (with $\cos\frac{m}{2}\theta$ and with $A = B$), whereby the function $f(\theta) = |\cos m\theta|^{1/n_4}$ fits into the superpolygon given by the denominator, the lower row of Fig. 5.7. Alternatively, $\varrho(\vartheta)$ is the *ratio* between $|\cos m\theta|^{1/n_4}$ and $\sqrt[n_1]{\left|\cos\left(\frac{m}{2}\vartheta\right)\right|^{n_2} \pm \left|\sin\left(\frac{m}{2}\vartheta\right)\right|^{n_3}}$. We say that the original function $f(\theta)$ is transformed, just like the circle with radius $f(\theta) = R$ is transformed into starfish or regular polygon. Hence we call Eq. 5.12 the Gielis Transformation.

Spirals are another logical choice of functions. There is a wide variety of spirals, with the Archimedean spiral $\rho(\vartheta) = a \cdot \vartheta$ and various conchoids. The logarithmic spiral $\rho(\vartheta) = e^{a\vartheta}$ is a most interesting one, because of its connection to the exponential function and the catenary, its origin in the hyperbola, and its ubiquity in natural shapes (Fig. 5.8). The famous *Nautilus* shell is one of the iconic and quintessential examples of mathematics in nature but with our transformations most of the shells of mollusks can be generated (Fig. 5.9).

There are a wide number of spirals in geometry. Particular examples are sine spirals, Cornu-Euler spirals, Archimedean and logarithmic spirals. Interestingly, Gino Loria wrote that a more general definition of spirals could encompass all curves for which the easiest and most natural parametrization is via polar coordinates: „*Vorläufig wollen wir als Spiralen alle Kurven bezeichnen, deren einfachste und geeignetste analytische Darstellung man bei der Anwendung vor Polarkoordinaten erhält*" [72].

Spirals may be interpreted very broadly, with a position vector fixed at the pole and turning around the axis perpendicular to this plane and through the pole, in which the length of the vector can change, uniformly or with certain speed variations, allowing for a purely geometrical study.

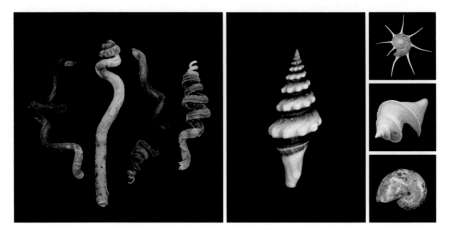

Fig. 5.9 Superspirals in nature. Copyright Violet Gielis

Fig. 5.10 Morphing a circle into a starfish. The value of $m = 5$ in all cases

Continuous Transformations

In 1917 D'Arcy Wenthworth Thompson published one of the finest books ever on the relationship between mathematics and biology, *On Growth and Form* [26]. One of the most important contributions in the book to biology was the transformation theory, in which he shows with simple diagrams how certain species of fish, the shape of leaves, or the skulls of mammals are connected geometrically. D'Arcy Thompson unveiled how the connections could be defined as continuous transformations. Likewise, many shapes can transform into various other shapes in a simple way in which the generic symmetry is defined by the above equations.

A further interesting observation is in relation to symmetry and symmetry breaking. The latter term has become very popular in science in the latter half of the 20th century and has become a fully accepted part of scientific terminology. A (hypothetical) spherical egg cell is polarized after fertilization and all subsequent cell divisions lead to a symmetrical/asymmetrical body, like the transformation of a spherical egg into a pentagonal (non-biological) starfish. There are however quite a number of conceptual difficulties with such symmetry breaking concept: The symmetry has to break "from infinite" (based on the circle as a polygon with infinite symmetry), to five. In this respect, we can pose the following question:

When comparing the radial symmetry of a starfish and a circle, which of the following statements is true?

- *The circle has a higher symmetry than a starfish.*
- *The circle has a lower symmetry than a starfish.*
- *The circle has the same symmetry than a starfish.*
- *All of the above statements are correct.*

All four statements are actually correct from our point of view. Using Eq. 5.8, we have a starfish for symmetry $m = 5$. The circle (our classical Euclidean circle) is obtained either as a zerogon (for $m = 0$) or as a polygon with infinitely many edges in the Archimedian tradition of approximating a circle by polygons with increasing number of sides. But in Eq. 5.9, for $A = B; n_2 = n_3 = 2$ the denominator becomes 1 for ANY value of the symmetry parameter m. So it can also be 5, the same as the symmetry of the starfish.

Any supershape (e.g. Table 5.1) can be transformed into a circle, for $A = B; n_2 = n_3 = 2$ and the circle can be morphed or transformed into any other supershape, by changing the parameters. With Eq. 5.8, a circle can be of any symmetry, including a pentagon, as long as $n_{2,3} = 2$. So moving from a circle to a pentagon and a starfish requires no change in symmetry at all—it is five in all cases —and changing of the exponents suffices (Fig. 5.10). Since the circle is one of the pentagons (for $n = 2$) no symmetry breaking needs to be called upon to precisely describe this transformation. Variations in starfish (developmental, intraspecific…) can be quantified precisely, in line with D'Arcy Thompson's transformations.

Symmetry is not written in stone in mathematics. One way of defining symmetry is the preservation of some measure related to some characteristic of some objects under the application of certain kinds of transformations to these objects [106, 107]. And, depending on which characteristics, objects or transformations, or even measurements one decides to use to compare shapes before and after the transformations has been applied, quite a number of different symmetries may be in agreement with this definition. So, from a rigid Euclidean point of view, a starfish and a circle are not the same (they are not congruent figures), but from our point of view, they are the same, just as circle and square became members of one family through Lamé's work. They become commensurable, and can be compared.

For the sake of the argument, the starfish in Fig. 5.11 may be different species, different individuals within a species or even may depict a sequence of development of one individual starfish, starting as a very small structure upper left, down to a fully grown starfish lower right. So we may compare individual starfish among species, within species or instances in time of the growth of one individual as continuous transformations. The latter view is important since we can understand and study one starfish as an individual, changing shape and size during development, but remaining the same individual in its travel through space-time. Likewise, in a broader view,

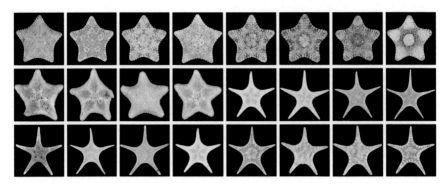

Fig. 5.11 Variations in starfish can be quantified precisely. Copyright Poppe Images

Fig. 5.12 Distances, area and polar moment of inertia I_p are conserved for changing m for fixed exponents (here all exponents $n = 1$)

differences between species, may be the result of the branching of an evolutionary tree, which becomes continuous, instead of discrete and contingent.

Invariance Properties of Supershapes

The equations also permit precise quantifications and then the description of a wide variety of shapes into one single equation gives many opportunities in understanding certain aspects of optimization in plants and natural shapes. The equations themselves permit direct calculation of associated characteristics such as area, perimeter, polar moment of inertia, defined by:

The **area** of a supershape

$$A = m \int_0^{2\frac{\pi}{m}} \rho(\theta)^2 d\theta \tag{5.13}$$

The **polar moment of inertia**:

$$I_p = m \int_0^{2\frac{\pi}{m}} \rho(\theta)^4 d\theta \tag{5.14}$$

The **circumference**:

$$s = m \int_0^{\frac{2\pi}{m}} \sqrt{\rho(\theta)^2 + \rho'(\theta)^2} \, d\theta \qquad (5.15)$$

We have the following invariances (for a reasonable choice of the parameters): The *invariance of length* $\rho(\theta)$ *at* $k\pi/m$ for all m, with k an integer. This follows directly from the equations. While in this expression $m = 0$ is formally excluded, the fact that distance is the same for any angle is the very characteristic of a circle, or zerogon ($m = 0$). Again, the parameter m defines how the points are spaced. With the introduction of $m/4$ the plane can be divided into any number of sectors, defined by m integer or rational. Equation 5.8 maps the values calculated for supercircles with $m = 4$, onto different axis systems defined by m.

For example, in a hexagon, the value at 60° ($2\pi/6$) is equal to the value of 90° in a square ($m = 4$ in Eqs. 5.8 or 5.9). The value at 30° in a hexagon ($\pi/6$) is equal to the value at 45° in a square. So distances at $k.2\pi/m$ are always the same, for any value of m and any real number k.

Second, *conservation of area for given exponents n, for all values of m*. Stated otherwise, when the rotational symmetry parameter m is changed, the shapes respond by making their sides more concave or convex, in order to preserve the area (Fig. 5.12). Third, is conservation of polar moment of inertia I_p (Eq. 5.14) for given exponents n, for all values of m.

Gielis Surfaces and Volumes

The curves in 2D-planes can readily be extended to surfaces. Instead of supercircles and superellipses we have "*supersurfaces, superspheres, superellipsoids* and *superquadrics*" [108]. The extension from polar to spherical coordinates is straightforward. Starting from $\left|\frac{x}{a}\right|^P + \left|\frac{y}{b}\right|^P + \left|\frac{z}{c}\right|^P = 1$ in spherical coordinates we obtain:

$$\left|\frac{r \sin\vartheta \cos\varphi}{a}\right|^P + \left|\frac{r \sin\vartheta \sin\varphi}{b}\right|^P + \left|\frac{r \cos\vartheta}{c}\right|^P = 1 \qquad (5.16)$$

Like in the two dimensional case this can be generalized, using real numbers, to [109]:

$$r(\vartheta, \varphi) = \frac{1}{\sqrt[n_1]{\left|\frac{\sin\left(\frac{m_1\vartheta}{4}\right)\cos\left(\frac{m_2\varphi}{4}\right)}{a}\right|^{n_2} + \left|\frac{\sin\left(\frac{m_1\vartheta}{4}\right)\sin\left(\frac{m_2\varphi}{4}\right)}{b}\right|^{n_3} + \left|\frac{\cos\left(\frac{m_1\vartheta}{4}\right)}{c}\right|^{n_4}}} \cdot \alpha(\vartheta, \varphi) \qquad (5.17)$$

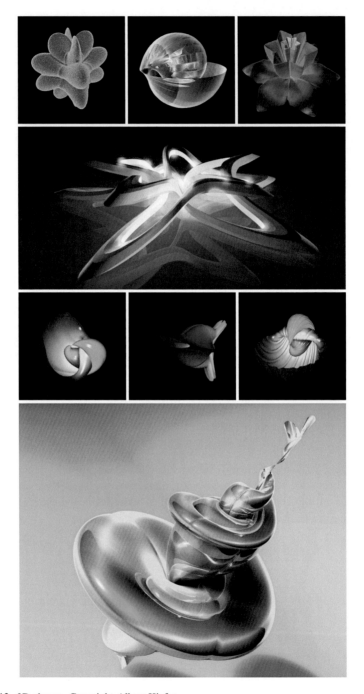

Fig. 5.13 3D shapes. Copyright Albert Kiefer

As transformations of functions, with $\alpha(\vartheta, \varphi) = 1$ the unit sphere, Eq. 5.17 can be generalized immediately to any dimension. Another way of defining 3D shapes is as spherical products. Superellipsoids are defined as the product of one full circle and one half-circle, both centered at the origin and perpendicular to one another, and defined by $m(\vartheta)$ and $h(\varphi)$:

$$m(\vartheta) = \begin{bmatrix} \cos \vartheta \\ \sin \vartheta \end{bmatrix} \quad \text{for} \quad -\frac{\pi}{2} < 0 < \frac{\pi}{2}, \quad \text{and} \quad h(\varphi) = \begin{bmatrix} \cos \varphi \\ \sin \varphi \end{bmatrix} \quad \text{for} \tag{5.18}$$
$$-\pi < 0 < \pi$$

A sphere of radius R is defined by the "*spherical product*" [96] of $m(\vartheta)$ and $h(\varphi)$:

$$\begin{cases} x = R \cos \vartheta . \cos \varphi \\ y = R \sin \vartheta . \cos \varphi \\ z = R \sin \varphi \end{cases} \tag{5.19}$$

In a kinematic way the sphere is a surface of rotation obtained by turning a circle along a semicircular path. In the very same way, a 3D-supershape can be defined as the spherical product of two superpolygons $\varrho_1(\vartheta), \varrho_2(\varphi)$ based on Eq. (5.8) or two 2D-supershapes $(\varrho_1(\vartheta), \varrho_2(\varphi))$ based on Eq. (5.12)

$$\begin{cases} x = \varrho_1(\vartheta) \cos \vartheta . \varrho_2(\varphi) \cos \varphi \\ y = \varrho_1(\vartheta) \sin \vartheta . \varrho_2(\varphi) \cos \varphi \\ z = \varrho_2(\varphi) \sin \varphi \end{cases} \tag{5.20}$$

Hence 3D-supershapes can be defined on the basis of two perpendicular sections $\varrho_1(\vartheta), \varrho_2(\varphi)$ (Table 5.2 [110–112]), The perpendicular sections can also be understood as follows: the left 2D shape of the pair is defined in the XY plane and is turned around the Z-axis following the path defined by the right 2D shape of the pair. This makes 3D shapes conceptually simple: 3D supershapes based on two perpendicular sections, in which one section is the *object* and the other section is the *path*.

Table 5.2 Spherical product of two superpolygons

Cross section $\rho_1(\theta)$ × cross section $\rho_2(\varphi)$ = 3D shape in (θ, φ)		
Circle	Circle	Sphere
Circle	Rectangle	Cylinder
Circle	Ellipse	Ellipsoid
Square	Square	Cube
Square	Rectangle	Beam
Square	Triangle	Pyramid
Circle	Triangle	Cone
Small circle	Large circle	Torus
Small square	Large square	Square torus
5/2 polygram	Circle	Bisshop's Cap

When both path and objects are superellipses (Eq. 5.4 with $m = 4$; $n_1 = n_2 = n_3$), superellipsoids are obtained. When both $\varrho_1(\vartheta)$ and $\varrho_2(\varphi)$ are zerogons or circles ($\varrho_1(\vartheta) = \varrho_2(\varphi) = 1$), a sphere results. When both $\varrho_1(\vartheta), \varrho_2(\varphi)$ are squares, the resulting supersurface is a cube. Combining a circle and a square results in a cylinder and so on. For $\varrho_1(\vartheta) < \varrho_2(\varphi)$ we get a torus. This torus can also be square if both path and object are square. When a twist is made we get Möbius bands. Inequalities define solids as well, the complete domain.

The possibilities are then infinite (Figs. 5.13 and 5.14). Instead of morphing spheres and turning square, we can also extrude a shape along a path perpendicular to that shape. A cylinder can be considered as a circle extruded along the vertical passing through its center; or as circles stacked onto each other; or as the spherical product of circle and rectangle. In fact all comes down to the same parametrization. In this case we have generalized cylinders, whereby the circles can morph into different cross sections at different heights as in cacti (Fig. 5.15). Depending on the content of water in the column, the cross section will change in cacti and this can be quantified in a precise way, within one single framework.

The shapes generated are surfaces using equality only. If however, inequalities are used, the inside of the shape is described as well. This allows for defining a certain width to the surface, or from a material point of view, to add specific homogeneous or non-homogeneous materials. The possibilities are then, indeed, without limits (Figs. 5.16 and 5.17).

Fig. 5.14 Fluid mechanics and patterns as supertori

Fig. 5.15 Cacti and stapeliads as generalized cylinders with supershaped cross sections

Generalized Möbius-Listing Surfaces and Bodies

The helices, tori and Möbius bands in space can have any number of twists, any cross section or even any way of avoiding selfintersection. They are toroidal knots with a twist. Such analytic representations or equations substitute for "recipes" or "(computer) algorithms" to "generate" Möbius strips, tori, helices or more complex shapes. They are no longer simple parametrizations of different shapes, but they combine these shapes into one equation, allowing for continuous transformations from circle or sphere into any other shape, including knots and certain polyhedrons, irrespective of topological constraints (Fig. 5.17).

In this sense transformations from sphere to torus e.g. with intermediates as ring, horn and spindle tori, become very natural. The reduction of "recipes" to equations, with a concomitant reduction of shape complexity, has always led to progress in science. This "old kind of science" has been true from Greek mathematics to the

Fig. 5.16 Future Fossils [113, 114]. Copyright Albert Kiefer

Fig. 5.17 Modern Architecture. Copyright Albert Kiefer

present day. These analytic representations provide tractable formulations for suitable studies in geometry (more particularly in low dimensional geometry and topology), applied mathematics and mathematical physics. Moreover, such general class of shapes is very natural indeed: It is kinematics and moving through space rotation and twisting is described by the Frenet-Serret-Pagani equations (G.M. Pagani was an Italian mathematician living in Belgium). One can generate an infinite number of shapes, based on kinematic or dynamic considerations.

Whereas our approach was one of discovery and understanding, the goal of any geometrical strategy should always be that in the proposed analytical representation, each function and each parameter in the formula has a geometric meaning. It turned out that in Tbilisi, 4000 km from my place Ilia Tavkelidze developed precisely this. Since the nineties he had been working on a concise and simple description of complex shapes, as compositions of elementary movements in the

sense of Gaspar Monge. It was through our mutual friend P.E. Ricci, that a natural
contact developed.

A wide class of geometric figures called "Generalized Twisting and Rotated"
GTR bodies are described by his analytic representation with many classical objects
(torus, helicoid, helix, Möbius strips). A subset are Möbius-Listing surfaces and
bodies [115, 116]. The original motivation here was not natural shapes, but was
dedicated to topological properties of surfaces and bodies, in search for a uniform
description and based on elementary movements rotations and translations. The
original works of Gaspard Monge on elementary movements date back two centuries, and 150 years separate the original description of Möbius-Listing ribbons
and the complete description of Generalized Möbius-Listing surfaces and bodies. In
science, good things need time.

These general developments were initiated independent from explorations of the
Gielis curves and surfaces in the study of natural shapes or from similar developments in computer graphics (Fig. 5.18). Combining both methods is most natural in
the same idea of object and path, using only elementary movements. Gielis curves
or surfaces as input for this general class expand this class of surfaces further to
bodies with interior structure [117, 118].

Generalized Möbius-Listing Bodies GML_m^n are defined by the analytic
representation:

$$
\begin{cases}
X(\tau,\psi,\theta) = \left[R(\theta) + p(\tau,\psi)\cos\left(\frac{n\theta}{m}\right) - q(\tau,\psi)\sin\left(\frac{n\theta}{m}\right)\right]\cos(\theta) \\
Y(\tau,\psi,\theta) = \left[R(\theta) + p(\tau,\psi)\cos\left(\frac{n\theta}{m}\right) - q(\tau,\psi)\sin\left(\frac{n\theta}{m}\right)\right]\sin(\theta) \\
Z(\tau,\psi,\theta) = \left[p(\tau,\psi)\sin\left(\frac{n\theta}{m}\right) + q(\tau,\psi)\cos\left(\frac{n\theta}{m}\right)\right]
\end{cases}
\tag{5.21}
$$

or, alternatively,

$$
\begin{cases}
X(\tau,\psi,\theta) = \left[R(\theta) + p(\tau,\psi)\cos\left(\psi + \frac{n\theta}{m}\right)\right]\cos(\theta) \\
Y(\tau,\psi,\theta) = \left[R(\theta) + p(\tau,\psi)\cos\left(\psi + \frac{n\theta}{m}\right)\right]\sin(\theta) \\
Z(\tau,\psi,\theta) = \left[p(\tau,\psi)\sin\left(\psi + \frac{n\theta}{m}\right)\right]
\end{cases}
\tag{5.22}
$$

where $p(\tau,\psi)$ is a given function which can define *m*-symmetric surfaces or bodies,
of which the ends can be connected with rotations or twists. Shorthand notation for
GML_m^n bodies has both upper and lower index. The lower index gives the symmetry
of the body or surface. So, a GML_6^{12} torus is a 6-symmetric cylinder with 12 twists
(obviously the twists are made relative to the symmetry of the figure).

The Möbius strip is an icon of mathematics. This "twisted cylinder" is obtained
by giving one end of a rectangular strip of paper a twist before joining both ends.
Simple as this may seem Möbius strips are one-sided, non-orientable surfaces with
some counterintuitive properties. Tracing out a path on a Möbius strip will show
that it has only one side and the whole strip can be painted in one color without

Fig. 5.18 Architon GML. Copyright Bert Beirinckx

lifting the paint brush. If some form of identification is done along the central line before joining, one will find that orientations are reversed after rejoining.

To cut a Möbius band either the cut is made along the basic line or along some other line, called s-line and depending on the number of twists given (even or odd), this gives different outcomes, either knots or links. On GML_m^n bodies or surfaces the B-line and s-lines can be drawn and also B-zones (zones containing the B-line) or s-zones (zones not containing the B-line) can be defined. In this group of bodies the restriction is given by the fact that the basic line is a circle. When GML_m^n bodies are cut along well-specified lines, intricate patterns result, with close connections to links and knots. In many cases the results of the cutting have been completely classified, and also the general case of convex cross sections has been proven. There is a direct connection to the shape and number of resulting bodies and zones defined by self-intersecting supershapes (Fig. 5.19).

One may be inclined to think that should be common knowledge, but in fact the systematic study of GML_m^n and twisted and rotated bodies has only started in the last two decades, also in differential geometry [119–121]. Applications are numerous, in biology, physics and chemistry. The heart itself is a folded Möbius band or GML_m^n and if one decreases the size (cone-like) one can study vortices (of water, light...) with inner structure. The shapes in Fig. 5.19 could be very complex dynamical systems, whereby the cutting or separation of zones is not a physical cutting with scissors, but could be a separation due to vibration and nodes caused by this vibration.

Many dynamical systems, which are considered chaotic today, may well have a rich structure with discernible knots and links structure. The composite structure may be traced back to the original GML_m^n with internal structure, and one could try to find the physical systems that led to the separation and the complex structure.

Fig. 5.19 Knots and links after cutting GML ribbons. Copyright Albert Kiefer

Fig. 5.20 Goethe's rose grown through

From a kinematic point of view, each of the components in Eqs. 5.21 and 5.22 has a precise geometrical meaning. In this way complex movements can be decomposed in simple movements [118].

Analytic Representations or Transformations?

Various names—supershapes, superpolygons, Gielis Formula *GF*, Gielis Superformula *GS* or Superformula *SF*—were used so far. The history of the name is interesting in itself. I originally called Eq. 5.8 the Superformula, as a generalization of superellipses and supercircles, names that are common names for Lamé curves. I used this name in the book *Inventing the Circle*, published in 2001 (October) in Dutch, in April 2003 in English. In April 2003 the first scientific paper was published as invited special paper in the *American Journal of Botany*. The names *Gielis curves, surfaces and transformations* were launched by L. Verstraelen at a Geometry meeting in Kragujevac (Serbia) in June 2004 [12]. Since then Superformula, Gielis Formula and Gielis Superformula have been used in many scientific articles referring to or using the Superformula.

To clarify the various names that are in use, we have to distinguish between *analytic representations* or *transformations*. Are the equations used, analytic representations, or do we consider them as transformations? In my opinion we can use both. I use *analytic representations* only in a mathematical sense, but *transformations* in a much wider sense.

When the Superformula is turned into equality or an equation (*GF* = some number or function) or inequality (*GF* smaller than or equal to some number or function) it becomes an *analytic representation* of actual shapes (even if infinitely many) for specific parameter choices (or ranges). Another word is parametrization. In the same way it represents surfaces (equality in 3D), volumes (inequality in 3D), or n-dimensional manifolds or submanifolds. We can define annuli and shells with different outer and inner boundary. Through the inequality, the domain is defined or parametrized as well.

So, in this sense analytic representation is the correct term, for describing supershapes or Gielis curves, -surfaces, -bodies, -lines, -manifolds and -submanifolds. GML_m^n bodies and surfaces are also analytic representations. The parametrizations can also contain functions, rather than parameters, whereby values of exponents change over time, or along a shape (as in cacti).

Besides this *synthesis* step, in which the input are numbers (or functions) it is also an analytic representation in (at least) two other meanings. First, when we do the inverse with *analysis* of shapes or patterns. We scan a shape and try to estimate the best parameter choice. It then provides the best Superformula analytic representation for a set of measured points. The formula is very general, while for modeling of natural shapes we often need less. For example, for the analytic representation of bamboo leaves or tree rings we need only two parameters. Second, when we try to find an optimal shape determined by a specific parameter set resulting from some *optimization* algorithm as for antennas, nanoshapes in chemistry, solar panels, wind turbines, gear etc. Hence it is an *analytic representation* or *parametrization* of specific shapes in those three ways (synthesis, analysis and optimization).

The original title of the paper in the American Journal of Botany is *A generic geometric transformation that unifies a wide range of natural and abstract shapes.* While the unification is the discovery of one generic, uniform analytic representation, it is also a transformation. Indeed, shapes can be easily transformed into one another by changing numbers, either in discrete steps or in continuous ways. So, while we understand that specific shapes can be described by their analytic representation, in this sense it is a transformation and also, obviously, geometric. So there is a historical reason to use *"transformations"*, and there are a number of other ways and meanings in which we can use the word transformation.

Natural shapes can change shape and during growth, or evolution, the shapes morph or transform into other shapes. Transformation has the same meaning here as "changing shape" or morphogenesis. I also want to stress that it is geometric and generic so that it can be used in any field of science, avoiding the restriction of botany or biology only. Morphing shapes is also the subject of topology. One way of viewing history is that topology was born because the appropriate geometrical formulae were lacking. We now have a transformation in a geometric-topological sense. It may not be as general yet as topological transformations, but general enough for most practical purposes. So, yes we can have square tori and we can transform these into a cup, or a sphere, shapes with other values for topological genus *g*.

The next step, by considering it as transformations in Euclidean space, adds to the wealth of Euclidean geometry. Since we can now make a vector stretch and turn along a square, we define a square rotation. It adds to Euclidean geometry (with translations, rotation, reflection, glide reflections and scaling), the idea of turning square. It extends Lamé's idea of coordinate systems adapted to the shape, of fitting a most natural coordinate systems to natural organisms or phenomena, embedded in Euclidean space. In the end, having geometries for natural shapes with our transformations we can understand that these geometries are subsets of a larger one.

In botany the word transformation even has a very special meaning, since Goethe wrote:

> Es war mir nämlich aufgegangen, dass in demjenigen Organ der Pflanze, welches wir als Blatt gewöhnlich anzusprechen pflegen, der wahre Proteus verborgen liege, der sich in allen Gestaltungen verstecken und offenbaren könne. Vorwärts mal rückwärts ist die Pflanze immer nur Blatt.... [122].

Everything is leaf and from this simple principle the greatest diversity or variety (*Mannigfaltigkeit*) of shapes in leaves, flowers and reproductive systems is possible. He was inspired by the phenomenon of proliferations in rose, where new vegetative shoots develop inside flowers (Fig. 5.20). Goethe's views *Alles ist Blatt* has been corroborated by plant molecular biology [123], and only for this we should consider Goethe as a great observer and a true scientist.

Goethe understands the *Mannigfaltigkeit* in nature as variations on a theme almost seventy years before Riemann introduced manifolds into geometry as the multi-dimensional generalization of surfaces and curves, and before Darwin's *"from so simple a beginning endless forms most beautiful and most wonderful have been, and are being, evolved*" [124]. Goethe provided new glasses to view nature as a continuum and thus inspired many of the developments in science and more, directly or indirectly. His views on colors, whereby he showed that there are many other colors than the distinct spectral colors of Newton, have become an integral part of the science of colors [125].

The word transformation is used in a very general way, both as concept and technique, that can be applied in many fields. Once more, it is a matter of words: stating that shapes and phenomena are unified by one transformation, is the same as saying that they can be described by the same analytic representation, of which parameters can change in time or space, or both. This distinction applies to all mathematical modeling.

Let us remember in this sense the wise words of René Thom: "*All models divide naturally in this way into two* a priori *distinct parts: one kinematic, whose aim is to parametrize the forms or the states of the process under consideration, and the other dynamic, describing the evolution in time of these forms*" [9]. There is a continuous transformation from circle to square, to starfish, eggs, supereggs, and many more shapes [126]. It is geometric topology, in which a precise geometric meaning can be assigned to any shape, the kinematic parametrization, and the way from one shape to the other, the dynamic one, in which time can be a parameter. Hence the names Gielis Formula, (Gielis) Superformula or Gielis Transformations are used interchangeably.

Part III
Διορισμός—Determinatio

Chapter 6
Pythagorean-Compact

> *What is essential for the topological complexity of an object is not the degree of the equation, but rather the number of monomials that appear in the polynomial with nonzero coefficients. Thus, we have the problem of "oligomials", i.e. the topology of objects specified by polynomials of arbitrarily large degree but with a restricted number of monomials.*
>
> Arnold and Oleinik [127]

A Zillion More

As generalization of Lamé curves, the Gielis formula is deeply rooted in a long tradition in mathematics of developing compact methods for shape description, striving for a uniform description. To each interesting development in science there is also a foreground, a future. While it is a powerful formula, opening doors in science and mathematics, we must not think of the Superformula or Gielis Formula as something ultimate. It generalizes Pythagoras, the Pythagorean Theorem, trigonometric functions, conic sections and more, but should be thought of as an inspiration to find and develop further generalizations.

It is a compact expression with circular trigonometric functions having angles as arguments, but we can consider the arguments of cosine and sine to be functions. Indeed, a generalization of the Superformula with elliptic functions has been proposed [128]:

$$\varrho(\vartheta) = c(\vartheta) \left[\left| \frac{1}{A} \cos \left(\Phi_{I,II}(\vartheta) \right) \right|^{n_2} \pm \left| \frac{1}{B} \sin \left(\Psi_{I,II}(\vartheta) \right) \right|^{n_3} \right]^{-\frac{1}{n_1}} \tag{6.1}$$

where $\Phi_{I,II}, \Psi_{I,II}$ are suitable amplitudes of Jacobi elliptic functions. This yields for example shapes of *Vinca* flowers (Fig. 6.1), but these can also be obtained using the superformula, by slightly shifting the angle in the denominator by a factor ε (Fig. 6.1).

© Atlantis Press and the author(s) 2017
J. Gielis, *The Geometrical Beauty of Plants*,
DOI 10.2991/978-94-6239-151-2_6

Fig. 6.1 *Vinca major*

$$\varrho(\vartheta) = \frac{1}{\sqrt[n_1]{\left|\frac{1}{A}\cos\left(\frac{m}{4}(\vartheta - \varepsilon)\right)\right|^{n_2} \pm \left|\frac{1}{B}\sin\left(\frac{m}{4}(\vartheta - \varepsilon)\right)\right|^{n_3}}} f(\vartheta) \qquad (6.2)$$

This generalization was introduced by R. Chacon: "*Since the price of the strikingly broad and complex diversity of natural shapes is the inherent nonlinear character of the processes giving rise to them, one should expect that even the most elemental formulas aimed at describing them should be expressed in terms of nonlinear mathematical functions closely related to general (i.e. nonlinear) natural processes*" [128]. The proposed Jacobian elliptic functions were solely intended as examples of this line of thinking.

Dealing with modeling geometrical shapes, however, this selection appears quite restrictive, since the non-linearity is restricted by the use of these particular functions. In order to generalize the Gielis formula, it is simpler to introduce the following more general equation [129]:

$$\varrho(\vartheta) = c(\vartheta) \left[\left| \frac{1}{A} \cos\frac{(m_1 f_1(\vartheta))}{4} \right|^{n_2} \pm \left| \frac{1}{B} \sin\frac{(m_2 f_2(\vartheta))}{4} \right|^{n_3} \right]^{-\frac{1}{n_1}} \qquad (6.3)$$

where $\vartheta \in [-\pi, \pi], f_1, f_2$ and $c(\vartheta)$ are continuous functions; m_1, m_2, A and B are real positive numbers and n_1, n_2, n_3 are integers or real numbers, for example:

$$f_1 = |3(\vartheta - \pi)| \quad \text{and} \quad f_2 = 3\cos 2(\vartheta + \pi)^2$$

In the original formula 5.8, we stretched (or shrunk) the radius for constant spacing of angle ϑ. The next step is to stretch or modify these angles and their spacing. The genius of Pythagoras was to separate variables avoiding geometric means between variables. In Eqs. 5.8 and 5.9, this idea is preserved and so is the structure of the formula. Generalized forms of the Gielis formula (Eq. 6.3) retain this very compact representation and structure (Figs. 6.2 and 6.3).

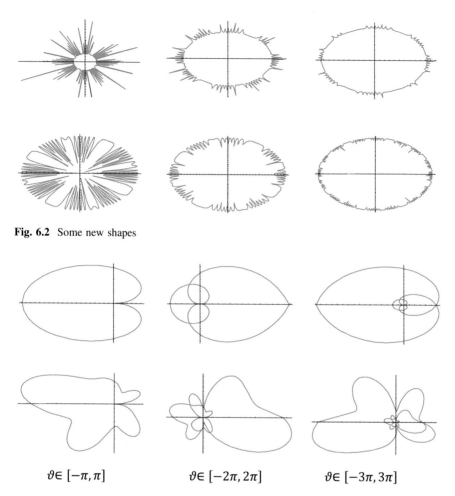

Fig. 6.2 Some new shapes

$\vartheta \in [-\pi, \pi]$ $\vartheta \in [-2\pi, 2\pi]$ $\vartheta \in [-3\pi, 3\pi]$

Fig. 6.3 New shapes with higher periodicity

Thus far we have assumed that the functions are $2k\pi$ periodic, defined by one single set of parameters. Of course, we can define curves in segments as well, with transitions between segments in any way we want. We may define a first supershape between 0° and 34°, a second one between 34° and 278° etc. It is furthermore easy to imagine that in these transitions, or along the complete shape, the exponents change as some functions of the angle. Another option is to replace the exponents by functions, defining change of time or space. Any of the parameters could be substituted by a particular function, so that the value of the parameters can change or morph along the shape itself. Between two different points on a shape other interpolation functions can be defined changing one particular distance in a different one. The possibilities are really infinite, while the basic structure of the formula remains compact and the same (Fig. 6.4).

Fig. 6.4 "Brilliant! Off the charts in terms of innovation" [130, 131]

k-Type Curves and a Generalization of Fourier Series

The shapes defined thus far are single shapes. There are several ways to make combinations and the simplest one is to simply sum shapes. Many well-known 2D curves are a sum of two shapes, e.g. the cardioid, defined in polar coordinates by $\rho(\vartheta) = 1 + \sin\vartheta$ or $\rho(\vartheta) = 1 + \cos\vartheta$. It has two parts, a base (unit) circle of radius 1 and the sine or cosine (the coordinate functions of this same unit circle). Modifications of the cardioid by our transformations describe leaves of various plants (Fig. 6.5) and basic leaf types (Fig. 6.6).

This can further be used as a starting point for building curves and surfaces as sums having the same center or origin. These sums are either of *infinite k-type*, for an infinite sum of shapes, or of *finite k-type* with *k* integer for partial sums. We refer to *k*-type Lamé-Gielis curves for a partial sum with *k* terms, with *k* a natural number. They are defined as the sum of *k* parts:

$$\varrho(\vartheta; f_i(\vartheta), A_i, B_i, m_{i1}, m_{i2} n_{i1}, n_{i2}, n_{i3}) = \sum_{i=0}^{k} \frac{1}{\sqrt[n_{i1}]{\left|\frac{1}{A_i}\cos\left(\frac{m_{i1}}{4}\vartheta\right)\right|^{n_{i2}} \pm \left|\frac{1}{B_i}\sin\left(\frac{m_{i2}}{4}\vartheta\right)\right|^{n_{i3}}}} f_i(\vartheta)$$

$$(6.4)$$

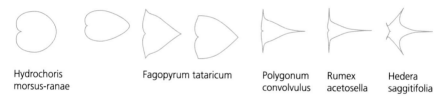

Hydrochoris morsus-ranae   Fagopyrum tataricum   Polygonum convolvulus   Rumex acetosella   Hedera saggitifolia

Fig. 6.5 Leaf shapes as transformations of a cardioid [132]

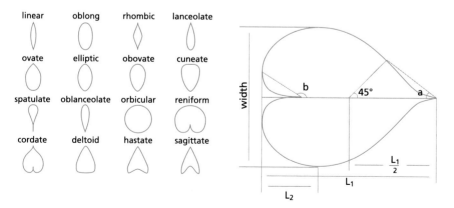

Fig. 6.6 Leaf shapes as variations on a superformular theme [133]

All two-dimensional normal-polar domains can be described or approximated accurately by selecting suitable modulator functions and parameters. Summation can be continued for any k. Adding three flower curves (or three modified Grandi curves) result in the outline of a flying bird, a 3-type curve. In this particular case the functions are cosine functions with arguments 3, 4 and 5 inscribed in a circle, a square and a pentagon, respectively. We can then construct sums of the shapes and their coordinate functions, whereby these coordinate functions can be inscribed in anisotropic spaces.

As a consequence all the 2D shapes discussed before (except the cardioid) are of 1-type. *Rhodonea* curves, also known as flower curves or Grandi curves, are also curves of 1-type, defined by $\rho(\vartheta) = a\sin(m\vartheta)$ or $\rho(\vartheta) = a\cos(m\vartheta)$ either with m or $2m$ petals depending on whether m is odd or even. Using $\rho(\vartheta) = \left|\cos\left(\frac{m}{2}\vartheta\right)\right|$ we arrive at m petals in flowers. Applying our transformations to the Grandi curves led to the flowers defined in Fig. 5.7. Another well known curve of 1-type is the lemniscate defined by $\rho(\vartheta) = a\sqrt{2\cos(2\vartheta)}$.

There is a natural alliance with trigonometric or Fourier series, which are defined in isotropic spaces. Now, each individual term of a Fourier series can be inscribed within supershapes, e.g. the flowers in Fig. 5.7. Each term of such series can be associated with different points of the shape space defined by Eq. 5.8 in which each point denotes one particular shape. One can thus define a Super-trigonometric or super-Fourier series (Eq. 6.5), or even a series of a more general type [134].

$$\rho(\theta) = \rho_0 a_0 + \sum_{k=1}^{\infty}\left(a_k\rho_k \cos\frac{m_k\theta}{4} + b_k\rho_k \sin\frac{m_k\theta}{4}\right) \tag{6.5}$$

Since the first constant term of the series a_0 can be associated with a particular transformation ρ_0 then any shape described by Eq. 5.8 or Eq. 5.9 is described precisely in only one term of this generalized series $\rho_0 a_0$.

Certainly, any of the above shapes could be described by for example Fourier series, but these are infinite by definition, and even shorter expansions are of considerable length. To quantify shape diversity (as in Begonia or maples) or even more complex shapes, elliptic Fourier methods are well suited and widely used. If, however, we need 200 harmonics to describe the outline of a leaf shape, the compact representation of cardioid leaves or petals of flowers as *1-type* curves, seems attractive. Especially if you know that the specific harmonics can differ widely among different leaves. In addition, we not only describe boundaries, but also the inside (disks or regions for self-intersecting polygrams).

In 3D, summing shapes can be performed comparable to Onaka's superquadrics, but with different position of the center. Shapes can then be combined via Boolean operations, introduced in computer graphics by Ricci in the 1970s [135]. We used those to develop a CAD modeler, to build individual shapes and combinations, like

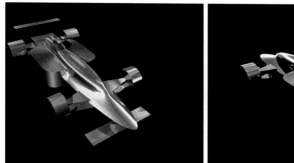

Fig. 6.7 A racecar in less than 900 bytes [139]. Copyright Albert Kiefer

planes, cars and trains. Such composite shapes could be encoded in extremely small file sizes, since only numbers need to be stored. Every point is defined uniquely (Fig. 6.7).

Pythagorean Trees, L-Systems and Fractals

A very interesting way to combine shapes with Boolean operations, while ensuring differentiability throughout the domain, is via R-functions, named after V.L. Rvachev (1924–2005). He showed that it is possible to translate Boolean operations into equations. In this way he could build complex structures or domains, to solve boundary value problems on these domains [136–138]. This gives clear advantages over a pure combination of shapes, for example over L-systems and fractals, two very powerful and popular systems for modeling of natural shapes.

L-systems are an algebraic way of modeling branching structures, named after Aristid Lindenmayer (1925–1989). In his landmark paper *Mathematical models for cellular interactions in development* [139], he introduced finite automata with possibility of multiplication. Finite automata had been introduced earlier by the mathematician John von Neumann (1903–1957), like Lindenmayer born Hungarian, but the extension with multiplication made it possible to model dividing cells. The system was later extended to model complete plants in *The Algorithmic Beauty of Plants* [140], and seashells in *The Algorithmic Beauty of Seashells* [141]. L-systems give only the skeleton of plants, but to this skeleton, tissues, leaves, flowers etc. can be added, to construct a complete plant (Fig. 6.8).

Fractals became a popular name through the work of Benoit Mandelbrot (1924–2010), who generalized earlier work of the French mathematicians Julia and Fatou. Fractals stressed scale-independence in nature, i.e. similar structures appear similar, irrespective of scale; zooming in or out does not fundamentally change the shapes [142]. Self-similarity is found throughout nature. The notion of non-integer or fractal dimension has a close connection to power laws; they are superparabolas. The notion of fractals combined many ideas that had been around for many years or even millennia. Scale-independence was well known earlier, with Pythagorean trees

Fig. 6.8 Architon in the Geometry Garden. Copyright Albert Kiefer

and spirals. The spiral of Theodorus of Cyrene shows the irrationality of roots of integers 3 up to 17 (except those that are squares themselves, namely 4, 9 and 16). In the Spiral of Theodorus of Cyrene, from vertex to vertex, the distance is always equal to 1, and the hypothenuses then form the series $\sqrt{1}, \sqrt{2}, \sqrt{3}, \sqrt{4}\ldots$ (Fig. 6.9).

Another visualization of the spiral is the Pythagorean Tree, a drawing made by J. Bosman in his book *The Wondrous World of Plane Geometry* [143] a decade before Mandelbrot. In this wonderful book, he builds upon the Pythagorean Theorem to draw the Pythagorean Tree as an example of scaling, self-similarity and series expansion. One could start from a isosceles triangle, or any triangle, and continue in any direction (Fig. 6.10). *"The area of the larger square equals the sum*

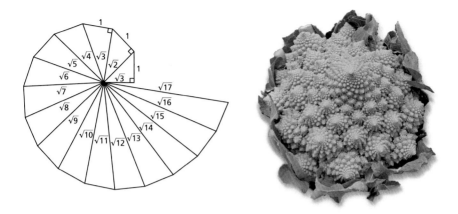

Fig. 6.9 Spiral of Theodorus (*left*) and Romanesco broccoli

Fig. 6.10 Pythagorean trees

of the two next ones, and is equal to the sum of the four further ones, etc. The *number of squares with common equal areas grows: 1, 2, 4, 8, 16, 32, 64, 128, 256, 528, 512, 1024, 2048.... into a wondrous construction of the Pythagorean tree made of sharp limit lines and the finest lace. Each square can be considered as the original square from which the tree developed, so that this tree consists of joining self-similar, ever smaller trees"* [143].

Fractals provided a new, powerful way of looking at shapes and phenomena and Benoit Mandelbrot's *The Fractal Geometry of Nature* [142] is simply a magnificent book. The system of arteries and lungs, or plant phyllotaxy in plants can be viewed as having fractal structure up to some degree (in nature that degree is not infinite as in mathematics). Fractals even inspired plant breeders and the Romanesco broccoli is a nice example of this. Cauliflowers resemble the Pythagorean tree in Fig. 6.10 left, while cutting Romanesco would give Fig. 6.10 right.

Fig. 6.11 Koch snowflake of order 3 (*left*) and 6 (*center*), and Sierpinski gasket. Copyright Luuk Verhoeven

Fractals may be considered as iterative systems, in which parts are added to a shape, or removed from a shape. Examples are the Koch snowflake, the Sierpinski Triangle and Cantor dust. In the case of the Koch snowflake on each side of a hexagon, even smaller hexagons are placed. On each side of these smaller hexagons, even smaller hexagons are placed. In Sierpinksi's Triangle one starts by deleting a smaller triangle can be deleted from the original one. These operations can be repeated ad infinitum. Starting from a line, one third can be removed, and from the two remaining one-thirds again the middle third part can be removed. After an infinite number of steps a fractal dust, Cantor's dust, results. L-systems can also be used to generate these fractals, as they operate on a string of letters, and string rewriting leads to rewriting systems. In this way many fractals can be generated, including Cantor dust, Sierpinski triangles and Koch snowflakes (Fig. 6.11).

Since fractals are often about applying a shape on the side of a basic shape, I cannot resist pointing out one of the most beautiful (but lesser known) theorems of Euclidean geometry. The Pythagorean Theorem (Euclid Book I, Theorem 47) is one of the highlights of Euclid's Books of Elements, and the application of squares to the sides of a rectangular triangle is used to prove it (used in Pythagorean trees or Theodorus's spiral). In Proposition 31 in Book V, Euclid shows that the shape we apply to the sides can be any shape. We can apply to each side of a rectangular triangle some regular or irregular shape (or supershape). As long as they have the same shape, the invariance property of the areas holds through the Pythagorean Theorem. Isaac Barrow: "*From this proposition you may learn to add or subtract any like figures, by the same method that is used in adding or subtracting of squares, in Scholium 47 in Book I*" [75].

R-Functions and Supershapes

For the composition of shapes we need a geometric language with an algebraic component. Whereas L-systems and fractals have as result a geometric shape, the road to get to this shape is algebraic and algorithmic. Indeed, in L-systems the algebra traces out only a skeleton to which flesh needs to be added. In fractals such as the Koch snowflake we are facing problems with differentiability. On the contrary, a geometric structure should immediately allow for applications of analysis and calculus. 'Geometric' means that from the outset the solution we are seeking allows to search for optimal solutions, in a purely geometric way. The development of natural objects can then be studied with natural curvature conditions defined from within geometry.

We can integrate both L-system-like structures and fractals into one coherent geometrical framework, using supershapes and R-functions. The language of L-systems and fractals is then enriched by providing an overall geometric structure, starting from geometry and ensuring differentiability at any level. Rvachev defined R-functions as geometrical translation of Boolean operations [138]. If we have two geometric domains we define the function as positive everywhere inside the domain, zero on its boundary and negative outside the domain. A union of these two domains then takes all positive values inside and the boundary of the composite shape has value zero everywhere. Rvachev succeeded in writing any Boolean function in a geometrical way (Fig. 6.12).

Consider a space of shapes in which individual shapes are represented by points in that space, the points are denoted by s_1, s_2, ... One point may represent a square of certain size, another point a triangle or a circle. He found that looking at such spaces in a geometrical way, he could transform logical Boolean operations between shapes into equations. At the same time various of the R-functions he defined (Boolean operations are a subset only), could guarantee differentiability up to any desired order. If two squares are combined, the transition between two shapes could be blended or smoothed.

The logical "conjunction" $A \cap B$ (or $s_1 \cap s_2$) denotes that a point has to belong to A and B at the same time. It is the intersection between two sets A and B. In the language of shapes: the point has to belong to both shapes. The logical "disjunction" $A \cup B$ ($s_1 \cup s_2$) denotes that a point has to belong to (at least) one of the sets. In the language of shapes: the point has to belong to either s_1 or s_2. We can use the triangle inequality and the law of cosines to derive R-functions (Eq. 6.6).

Both disjunction and conjunction can be represented by the R-function $R_\alpha(s_1, s_2)$. R_α functions are defined by:

$$R_\alpha(s_1, s_2) = \frac{1}{1+\alpha}\left((s_1 + s_2) \pm \sqrt{s_1^2 + s_2^2 - 2\alpha(s_1, s_2)s_1 s_2}\right), \qquad (6.6)$$

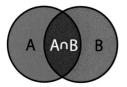 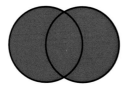 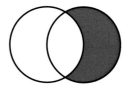

Fig. 6.12 Conjunction (*left*); disjunction (*center*) and relative complement of two sets

whereby $\alpha(s_1, s_2)$ is an arbitrary function such that $-1 < \alpha(s_1, s_2) < 1$. Setting $\alpha = 1$ leads to the functions $\max(s_1, s_2)$ or $\min(s_1, s_2)$ for $+$ or $-$ in Eq. 6.7, respectively

$$
\begin{aligned}
R_{\alpha=1}(s_1, s_2) &= \frac{1}{2}\left((s_1 + s_2) \pm \sqrt{s_1^2 + s_2^2 - 2s_1 s_2}\right) \\
&= \frac{1}{2}\left((s_1 + s_2) \pm \sqrt{(s_1 - s_2)^2}\right)
\end{aligned}
\tag{6.7}
$$

This can be considered by very easy geometrical considerations; for example, $\frac{1}{2}(s_1 + s_2)$ is the arithmetic mean. Adding or subtracting $\sqrt{s_1^2 + s_2^2}$ via these well-defined logical operations in the language of R-functions with $\alpha = 0$ is equivalent to a rectangular triangle (Eqs. 6.8 and 6.9). A remarkable property is that these functions are analytic everywhere, except in the origin.

$$
R_{\alpha=0}(s_1, s_2) = \left((s_1 + s_2) - \sqrt{s_1^2 + s_2^2}\right)
\tag{6.8}
$$

$$
R_{\alpha=0}(s_1, s_2) = \left((s_1 + s_2) + \sqrt{s_1^2 + s_2^2}\right)
\tag{6.9}
$$

This can be extended to three dimensions, with conjunction, disjunction or subtractions for ellipsoids and cubes, for spheres and cylinders. The Fichera domain, named after the mathematician Gaetano Fichera (1922–1996), consists of a cube from which a cube (1/8 of the original) is removed (Fig. 6.13). This also can be extended to more than two shapes, of course. In this way composite domains can be modeled efficiently, also fractals. In each step the shapes may remain the same or differ. The combination of course depends on the proximity of shapes. They can overlap or remain separate domains. In the latter case they are called disjoint.

Combining domains is only the beginning. The rationale for R-functions is to solve differential equations on domains where problems with differentiability arise, such as in corners of squares and polygons, or at cusps. This is the geometric

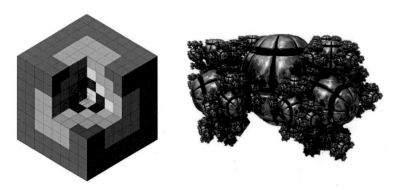

Fig. 6.13 Fichera domain (*left*) and superfractal (*right*)

motivation we need for a geometric algebra. The R-functions allow for any order of differentiability to shapes and their combinations (for positive integers m and p):

$$R_0^m(s_1, s_2) = (s_1 + s_2 \pm \sqrt{s_1^2 + s_2^2})(s_1^2 + s_2^2)^{\frac{m}{2}} \tag{6.10}$$

$$R_p(s_1 \vee s_2) = (s_1 + s_2) + (|s_1|^p + |s_2|^p)^{\frac{1}{p}} \tag{6.11}$$

$$R_p(s_1 \wedge s_2) = (s_1 + s_2) - (|s_1|^p + |s_2|^p)^{\frac{1}{p}} \tag{6.12}$$

$$R_p(s_1 \sim s_2) = \frac{(s_1 s_2)}{(|s_1|^p + |s_2|^p)^{\frac{1}{p}}} \tag{6.13}$$

This is a small selection of the possible R-functions, but the connection to Lamé curves becomes immediately clear from this definition, both the $(s_1^p + s_2^p)^{\frac{1}{p}}$ part and the $(s_1 + s_2)$ part with $p = 1$: to a linear combination of two shapes s_1 and s_2, a Lamé part is added or subtracted. Also $(s_1^2 + s_2^2)^{\frac{m}{2}}$ in R_0^m functions is a Lamé curve. In fact, the correspondence between R-functions with Lamé and Gielis curves has been formalized recently, since R-functions have been extended for p or m from integers to any positive value >1 [144].

In this way, differentiability is ensured up to order p or m, also non-integer values. Directly differentiating R-functions in the R_p system gives for the first partial derivative:

$$\frac{\partial f}{\partial x_i} = 1 \pm \frac{x_i^{p-1}}{(x_1^p + x_2^p)^{\frac{p-1}{p}}}, \quad i = 1, 2, \ldots \tag{6.14}$$

The order m or p determines the number of derivatives we can have. In Fig. 6.14, a 'rounding of corners' of a square is achieved. In this way a field can be defined whereby the function is positive inside the shape and negative outside, with the original shape for $f = 0$. Such fields can be colors, heat maps, or anything, just depending on how you define a field.

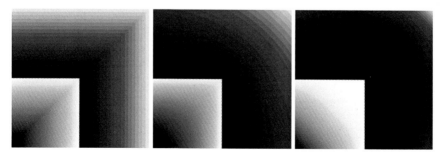

Fig. 6.14 $R_{\alpha=1}(x, y)$, $R_{p=2}(x, y)$ and $R_{m=2}^0(x, y)$ for R-disjunction. All have a constant sign in each quadrant [144]. Copyright Yohan Fougerolle

Complex Shapes and Self-intersections

One of the major advantages of R-functions is that we have a unified description for
shapes with our transformation. Any point in the space of supershapes has one
particular set of values namely $s_1(\vartheta; f(\vartheta), m, A, B, n_1, n_2, n_3)$. The inequalities that
describe the supershape and its interior can be used directly in R-functions for
combinations. Yohan Fougerolle was the first to make this fruitful combination [145,
146]. The rear axis of a truck in Fig. 6.15 is constructed out of 256 different
supershapes, and the turbine out of 5 supershapes. Notwithstanding the complexity it
is one single equation. Moreover, not only the points on the surface, but any point
inside the shapes is defined as well using the R-functions. One can see immediately
how this can be applied to plants and their modular structure. Phytomers, the basic
modules consisting of a stem, bud and leaf structure can be combined in a modular
way.

What about our self-intersecting polygrams? The power of R-functions becomes
fully apparent when we consider self-intersecting supershapes for rational Gielis
curves $m = \frac{p}{q}$, with p and q positive integers [147]. A ray drawn from the center will
cross the self-intersecting shape in more than one point. It is a multivalued function.
In Fig. 6.16 below an RGC curve with $m = 7/5$ has at most 5 intersections between
the curve and a ray originating from the centre, resulting in five identifiable layers.

In Fig. 6.16 the potential fields are shown for the conjunction between the outer
envelope and the complement of its inner envelope for RGC with $m = 5/2$; $m = 5/3$;
$m = 5/4$ based on a distance defined by the R-function used. Our original distance
functions are converted into fields and such fields could also be thought of as heat
maps, whereby more or less activity occurs in well-defined sectors. As a matter of

Fig. 6.15 Rear axis of truck
all in supershapes, and turbine
of windmill with cross section
displaying the potential fields.
Copyright Yohan Fougerolle

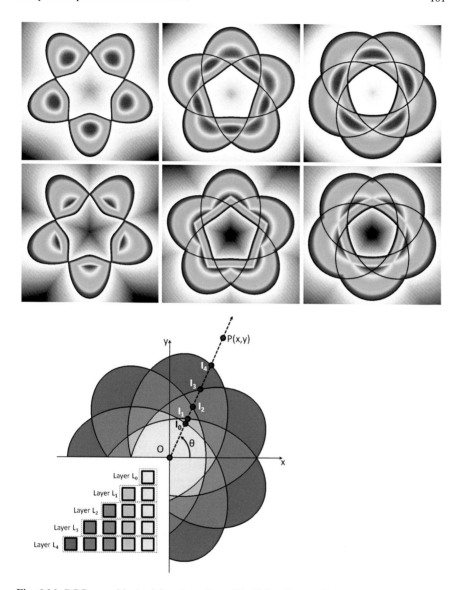

Fig. 6.16 RGC as multivalued functions. Copyright Yohan Fougerolle

fact, using the appropriate combinations and R-functions, each of the sectors can be studied independently from every other sector with some examples shown in Fig. 6.17. This strategy can be very useful. We can relate the sectors in Fig. 6.17 one-to-one on sectors and shapes resulting from a cutting process in Generalized Möbius-Listing bodies. The heat maps of Fig. 6.16 might relate to increased

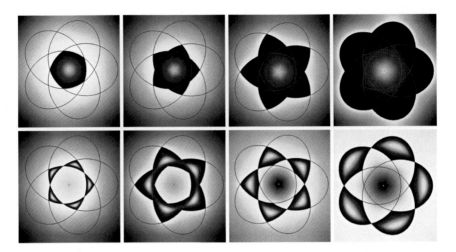

Fig. 6.17 Examples of the Boolean differences between consecutive layers. Copyright Yohan Fougerolle

activity of genes, and the development of floral organs inside whorls with spiral origin [147].

In the same way we can define fractals (which we limit here to polygon-like structures). A snowflake of Koch in three and six steps and a Sierpinski triangle can be constructed using supershapes combined via R-functions (Fig. 6.11) [148]. Not only are these shapes captured in a single equation, but also here potential fields can be defined. A main difference with fractals or L-systems is that these are not rewriting rules, replacements or substitutions. This allows for defining fields in the same way as above. The blue plane cuts (Fig. 6.11) through the whole structure exactly for $f = 0$. Gielis transformations and Rvachev functions form a very natural alliance [144], for regular and irregular curves, in the sense of [149].

R-Functions and Co-bordism

It was noted that a union between disjoint sets is possible as well, and in a certain sense any smaller or larger triangle in the Sierpinski gasket is disjoint from any other triangle in the zero level set. We can do the same for trees. If we start from two overlapping circles (or supershapes), and move them gradually apart, at a certain point they will separate. The circles become disjoint. However, they are still connected through their *"history"*. This history can be visualized as a tree, which branches at a certain point. In biology we know many trees: as plant models, as evolutionary trees, as developmental steps…

We can develop plant models in this way. Starting from a central column, the stem, branches will occur at well-defined points, depending on the species and

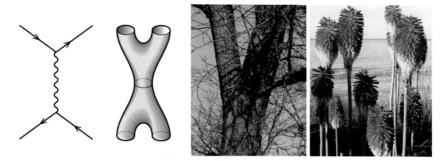

Fig. 6.18 Strings and real things

growth habit of plants, which can be tree-, shrub- or cactus like. And those three models are only three among many. Branching structures can include fasciated stems or other plant parts, abnormally grown together. Also here we have a beautiful connection to geometry and algebra. In the notion of *cobordism* [150], two disjoint sets make up the boundary of a manifold of one dimension higher.

Two manifolds of dimension n are called *cobordant* if their disjoint union is the boundary of a manifold of one dimension higher (For manifolds compactness etc. is guaranteed). We can continue such process further. If we make a horizontal cut through a tree, at the base we have one circle, while higher up we can have two or more circles in the same plane (note that the circles may actually be more elliptical, since the orientation of the branches is generally not vertical but oblique). These connections and cobordisms are also crucial in certain theories in elementary particles [151] (Fig. 6.18).

These manifolds lack inner structure and may be considered as simplified versions of biological structures like trees, in which the history is engrained in the wood and the specific shape of trees and branches. We can consider the tree and all of its sections as a manifold of which all the sections belong to the tree, obvious for any biologist. One can also consider the rhytidome (the bark) as a manifold (obviously of a less continuous nature), and boundary to the "tree-manifold". In nature, trees grow and add tree rings with every year of growth and a cross section will give lots of information about the tree. These can of course be considered as a distance function, and the tree rings as a "heat map" in the sense of *R*-functions. Each ring can be singled out as a zone (Fig. 6.19).

This history is very important in biology. We can represent a branching tree or root system, a rhizomatic growth form like bamboo or grasses, and evolutionary trees in this same way. Even more, since plants are clonal organisms, all clones that are generated from one single or a few phytomers, share a common descent and history. It is this history that connects millions of bamboos propagated via plant tissue culture and distributed all over the world, in gardens and plantations. The shapes resulting from cutting Generalized Möbius-Listing bodies with interior geometry also share this common history, and remain connected as disjoint sets through their history.

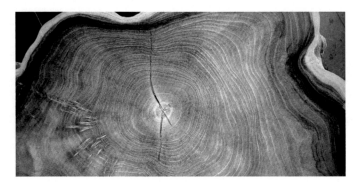

Fig. 6.19 Tree rings in teak. The central ones are Lamé curves

Shape Complexity and Oligomials

With our continuous transformations, we can morph any shape into many other shapes, even into combinations of shapes. This is not the classical way of thinking, where shapes are compared by their differences (shape, measure or topology). Moreover, we can morph simple shapes into extremely complex ones, but the compact representation remains the same. The shapes in Fig. 6.20 are not more complex than the simple shapes, just a different set of numbers. How then to define complexity?

If we want to determine the complexity of surfaces and curves, one idea is the degree of the polynomial describing the curve. Traditionally this led to linear, quadratic, cubic, quartic, quintic etc. curves. $6x^5 + 678x^4 = 0$ is a quintic curve of degree 5 in one variable. The degree is higher than that of conic sections, with degree 2, and more complex than quadratic ones. These can be solved in radicals, whereas not all quintics are solvable.

On the other hand, a superparabola of degree 678 $(y = x^{678})$ is of the same complexity as the classical parabola, just one number difference, and another graph. One can morph one into the other by changing the exponent from 678 to 2, continuous or discontinuous. Likewise, a Lamé curve with exponent $n = 567$ is not more complex than a circle with $n = 2$, from a "defining" point of view, nor is any of the shapes in (Fig. 6.20) more complex than sphere or torus. If we wish to describe the complexity of an object, it is not only the degree, but also the number of monomials in the polynomials, which describe the curve or object. In the case of Lamé curves, and the subset (with absolute values) the superellipses defined by $\left|\frac{x}{A}\right|^n + \left|\frac{y}{B}\right|^n = 1$, irrespective of the degree, the number of monomials is one in each variable. Gielis curves and surfaces generalize Lamé's ideas for any symmetry (integer and non-integer) and for any degree (including non-integer powers and roots).

Fig. 6.20 Supershapes with same complexity as a torus. Copyright Albert Kiefer

Such finite and short expansions can be named oligomials, a complete description in a few (oligo) terms. The name oligomials first appears in a text by V. Arnold and O. Oleinik, where they expressed the need for a better way to describe shape. *"What is essential for the topological complexity of an object is not the degree of the equation, but rather the number of monomials that appear in the polynomial with nonzero coefficients. Thus, we have the problem of oligomials, i.e. the topology of objects specified by polynomials of arbitrarily large degree but with a restricted number of monomials"*.

Fig. 6.21 Different shapes of
leaves of maples, *Hydrangea
petiolaris* and *H. quercifolia*

Oligomials relate directly to the topology of the objects; they can be specified by polynomials of arbitrarily large degree but with a restricted number of monomials. Moreover, as in the case of Lamé and Gielis curves these monomials are single variables, not geometric means. So we define complexity of shapes not in terms of computability, but based on its pure mathematical expression, and hence arrive at a suitable definition of geometric or topological complexity. And then our curves are topologically simple because of their oligomial structure.

All Gielis Curves Are Equally Simple

Lamé, *Rhodonea* and many other curves and surfaces, all date back to the 18th and 19th century. In the history of science we have two opposite tendencies of shape description. On the one hand we have the compact representations (Pythagoras, conic sections, Lamé-Gielis, Grandi,….). They allowed for transforming geometrical questions into simple equations (inspired by conic sections). On the other hand, we had the development of infinite series, such as Taylor and Fourier series. For practical purposes such infinite series are then truncated and referred to e.g. as Incomplete Bessel functions. This is an old tradition in mathematics, since it is known that Fourier series are modern versions of the computational techniques used by Ptolemy, as noted by a 19th century Italian astronomer, Giovanni Schiaparelli [152].

Our goal is to define curves as compact as possible, either as a modern version of the Pythagorean Theorem, or as a combination of a limited number of monomials. Indeed, *k-type* curves are finite polynomials (or monomials), combining the power of a compact and precise description, in one or a few terms, with the power of series expansions. The background to defining *k-type* curves is to develop finite type curves, comparable to Fourier series [153]. Describing a wide variety of shapes has hitherto been an intractable problem without resorting to infinite series. An immediate example: elliptic Fourier analysis allows for the description of very complex shapes, including plant leaves (Fig. 6.20). One of the advantages of Fourier is that it can handle very complex shapes including leaves [154].

The idea of *k-types* is based on *k-finite* type curves of Bang-yen Chen (in the framework of submanifold theory). His *k-type* curves are of infinite or of finite type, depending on whether their Fourier expansion with respect to arc length is infinite or finite [155]. From this geometrical perspective, he proved that there is one and only one closed curve that can be expressed in a finite Fourier series with respect to arc length, and that is the circle itself. In fact, this theorem implies that the circle is the only closed planar curve that is of finite type, namely of 1-type ($1T$) with all cosine and sine terms equal to zero, but any other curve necessarily has a Fourier expansion of infinite type (∞T). An alternative interpretation is that *all curves other than the circle, including the ellipses, are equally complex:* once their Fourier expansion starts, it never stops [156]. Their expansion contains infinitely many terms (∞T). Even the 200 harmonics to describe leaf shapes are a short version of the actual expansion.

Instead of infinite series, however, truncated for practical reasons, direct description of shape with Gielis curves provides a finite approach. This can be done in the compact Pythagorean structure of Generalized Fourier series. When Gielis curves are used as unit circles in a generalized Fourier series, i.e., on any term of a classical Fourier series a Gielis transformation can act, any Gielis curve of $1T$ is encoded directly, in one term only. They are of one-type ($1T$) and their expansion, once it starts, stops immediately, since all cosine and sine terms are zero. Hence, all Gielis curves, including the circle and Lamé curves, are equally simple; *k-type* curves (with $k > 1$) are almost equally *simple*, in one or a few terms. Complexity reduces to simplexity.

Pythagorean-Compact and Pythagorean-Simple

One further way to reduce complexity to simplicity is to introduce the precise notion of *Pythagorean-compact* description. The circle has its most compact expression in polar coordinates. All circles can be represented by $r = R$; all circles are defined by one number R which determines size, shape is fixed. With Pythagorean-compact we mean the description of size and shape in the same compactness of the Pythagorean Theorem, the most compact way of representing circles [129]. The expression $r = R$ however hides one important term, namely the Gielis transformation for $n = 2$ and $m = 4$. In this case $\varrho(\vartheta)$ is not a radius with variable length dependent on the angle, but fixed since this term is equal to one.

$$\varrho(\vartheta) = \frac{1}{\sqrt{(|\cos \vartheta|)^2 + (|\sin \vartheta|)^2}} \cdot R \quad \text{or} \quad \frac{\varrho(\vartheta)}{R} = 1 \qquad (6.15)$$

Supershapes and Lamé curves in our polar representation clearly have the same compact structure as the Pythagorean Theorem, with two perpendicular ways of measuring spacing (cosine and sine), and size parameters R or A and B in terms of ellipses. The main difference is the way of spacing the coordinates, governed by the parameters $\frac{m}{4}\vartheta$ and the exponents:

$$\varrho(\vartheta) = \frac{R}{\sqrt[n_1]{\left|\cos\left(\frac{m}{4}\vartheta\right)\right|^{n_2} + \left|\sin\left(\frac{m}{4}\vartheta\right)\right|^{n_3}}}, \quad \text{or}$$

$$\frac{\varrho(\vartheta)}{R} = \frac{1}{\sqrt[n_1]{\left|\cos\left(\frac{m}{4}\vartheta\right)\right|^{n_2} + \left|\sin\left(\frac{m}{4}\vartheta\right)\right|^{n_3}}} \qquad (6.16)$$

Hence, the Gielis Formula is Pythagorean-compact, and our transformations combine the compactness of the Pythagorean Theorem with its greatest advantage, the orthogonality (independence) of its variables. Within the same structure and same compactness we can define stretched radii and different spacing of angles, when $\frac{m}{4}\vartheta$ is generalized to some $f(\vartheta)$. So our formulas are topologically simple, Pythagorean compact and Pythagorean simple. 2500 years after Pythagoras, we can use the same structure not only to define circles but any supershape, abstract or natural.

Chapter 7
Generalized Intrinsic and Extrinsic Lengths in Submanifolds

The search for differences or fundamental contrasts between the phenomena of organic or inorganic, of animate or inanimate things, has occupied many men's minds, while the search for community of principles or essential similitudes has been pursued by few.

D'Arcy Wentworth Thompson

Embedded and Immersed

Lamé curves are not some strange specimen in the field of mathematics. Considering them as pure numbers—the outermost entries of any row of Pascal's Triangle—they are dual to the common "geometric mean" approach. Moreover, Lamé-Gielis curves and surfaces are excellent methods for describing natural shapes, to any degree of precision required. With this uniform description we have found a way of making shapes commensurable with one way of measuring, allowing us to compare a wide variety of abstract and natural shapes.

In this chapter we provide further examples of how Lamé curves and the Gielis Formula are fully immersed and embedded in the heart of mathematics, with clear links to some of the most well known applications of mathematics. It is surprising that it took so long to discover, since everything that was needed to develop the Gielis formula was ready in the first decades of the 19th century; not only ready in the sense that the ingredients were available, but really ready. Had Gauss and Lamé joined forces in 1818, the year of seminal publications of both, most of the results and combinations described in this book would have been known already two centuries ago.

Rigid and Semi-rigid Geometry

Gielis transformations find their origin in botany, defining unit circles in biology adapted to the shape. If we wish to study natural shapes and phenomena, it is necessary to agree on some basic definitions and assumptions, many of which may

© Atlantis Press and the author(s) 2017
J. Gielis, *The Geometrical Beauty of Plants*,
DOI 10.2991/978-94-6239-151-2_7

be very different from what we learned. Our Euclidean geometry is based on a Euclidean circle, with all points on the circle at a fixed distance from a fixed center. As humans we are free to roam in almost any direction (isotropy), and hence we can easily deduce that if we walk in any direction we will arrive after the same interval of time ("as the crow flies"), at points that form a circle. This rigid Euclidean circle is based on our own observations and perceptions of the world in which we are free to move, at least ideally [157]. In fact, our current laws in physics are based on isotropy.

As we know now, we have one transformation to describe many "anisotropic" shapes as transformation of a unit circle. In our approach of transformations, the notions of *unit circle* and *rigidity of rulers* play a crucial role. The notion of unit circle or unit ball in nature can be very different from our classical Euclidean circle. It is well understood in crystallography where the so-called Wulff shape provides the unit ball for a given crystallographic configuration, taking into account the molecular structure and the directions of growth, like for example the snowflake. Such unit balls introduce a natural anisotropy, with preferred directions of crystal growth.

So, why do we call these shapes unit circles or unit balls, if the definition of a circle is a "fixed distance" away from a center? This is simply because our classic circle is defined in an isotropic way, with the fixed distance as the crow flies (in all directions). To make this clear we can introduce the intuitive notion of semi-rigid rulers. To draw a Euclidean circle, we keep a piece of chord stretched from the center and trace out the circle. How do we draw a triangle then, a square, or any supershape? The answer is simple: you take an *elastic* band, fix it at a center and start drawing. The key is to stretch the elastic band more, or less, in certain directions. Then you can draw a starfish or a pentagon [1].

How to stretch it in a certain direction depends on the precise values of our transformations. We get another notion of distance from the center. In a very controlled and precise way our elastic band becomes a semi-rigid ruler and in the same way it is possible to construct a compass to trace out starfish or pentagons. I use the notion of semi-rigid to distinguish it from completely elastic as in topology. Squares, circles and starfish are topologically the same, but geometrically distinct. Our (continuous) transformations provide a bridge between geometry and topology.

Now, we divide our elastic or semi-rigid ruler into ten equally spaced parts and mark these with pegs (mathematically, we can imagine dividing an abstract line segment into as many equal parts as we want). If you now would count the number of pegs at $0°$ and $45°$ or at any angle you will always count the same number of pegs, namely ten, whether you draw a circle, a square or a starfish. This is obvious because when the rulers stretch, so will its subdivisions, as shown in Fig. 7.1. The number of pegs is always the same, irrespective of the angle and the endpoints will always be a fixed number of subdivisions away from the center and hence, by this definition they lie on a circle.

This then provides an answer to a problem posed by Richard Feynman: "*The difference between being a circle and being nearly a circle is not a small difference. It is a fundamental change so far as the mind is concerned. There is a sign of*

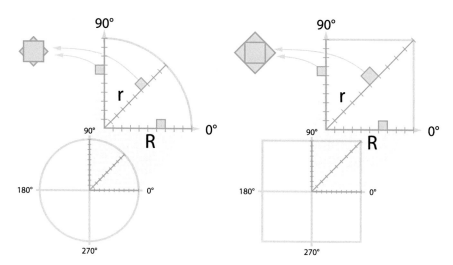

Fig. 7.1 Unit circles based on rigid and semi-rigid rulers. For illustrative purposes the observer is depicted as a two-dimensional square, but one has to consider the one-dimensional line, originating from a common center for each angle

perfection and symmetry in a circle that is not there the moment the circle is slightly off. That is the end of it; it is no longer symmetrical. Then the question is, why it is only nearly a circle? That is a much more difficult question. So our problem is to explain where symmetry comes from: Why is Nature so nearly symmetrical? No one has any idea why" [47].

The symmetry and the perfect object circle that Feynman refers to, find their origin in our ability to measure. From the earliest beginnings humans have chosen to measure with fixed distances, based on the rigid Euclidean circle. Nature on the other hand works with semi-rigid rulers and allows any of its shapes or patterns to have their own geometry. It is possible with our transformations to go from a circle to a square or to a starfish (and back) in a continuous way, and all such shapes will remain unit circles. Since it is continuous, we can describe the smallest possible changes, to discern among many very similar natural objects, like leaves on a tree. With our transformations Nature remains fully symmetrical, not nearly symmetrical.

Lorentz Transformations and SRT

If the "one-dimensional" observer on our elastic ruler would be you, then you would not even notice that the ruler stretches or shortens, because your whole world (the one dimensional line) will stretch or shorten, including yourself (Fig. 7.1). If you, as the observer, would count the number of subdivisions, you would find that at any angle, the intrinsic distances in your world remain constant. The observer in Fig. 7.1, who is precisely one subdivision in size, will count precisely ten subdivision in any direction, at all times.

We have two different viewpoints. From the classic Euclidean point of view, these transformations define a unit circle as a deformation of the Euclidean unit circle in which the radius is stretched or shortened. We can compute the stretching (lengthening or shortening) of the radius from the associated transformations. From the "intrinsic viewpoint" however, there is no stretching at all, for the observer on the radius. Most importantly, neither of the two has a preference; both are valid ways of measuring. If we insist on the Euclidean "fixed" ruler, we hold on to absolute definitions of distance and time. But they are a choice, not absolute.

In Special Relativity Theory a traveller A moving at high speeds will observe or measure nothing special, while a static observer B will observe and measure changes in A, and vice versa. Neither a fixed ruler nor a stretched (elastic) one can be considered as the "true measurement", which is exactly the same situation as described above, and it is in this sense that our formula can be considered as a general framework for the Special Relativity Theory.

In Special Relativity Theory the Lorentz transformation tells how space and time dilate and contract when one moves at uniform speeds. It is given by $\frac{1}{\sqrt{1-\frac{v^2}{c^2}}}$, with c the speed of light. Before Einstein this transformation was used by physicists to connect measurements with theory. Actually for Lorentz the transformed coordinates and fields were mathematical aids to facilitate the solution of differential equations that involved the wave operator. Henri Poincaré used ε^2 in $\rho(\vartheta) = \frac{1}{\sqrt{1-\varepsilon^2}}$ in his studies [158; 159], also a number between 0 and 1, just like $\frac{v^2}{c^2}$. The Lorentz and Poincaré transformations are based on a uniform motion along a line, but for moving into isotropic spaces, trigonometric functions can be chosen.

This transformation may be considered as a special case of Eq. 5.9 in the following way(s): for all exponents $n_{1,2,3} = 2$, $A = B = 1$, $m = 4$ and subtraction (instead of addition), we obtain $\rho(\vartheta) = \frac{1}{\sqrt{(\cos\theta)^2-(\sin\vartheta)^2}}$. For $n_2 = 0$, $n_{1,3} = 2$, $A = B = 1$, $m = 4$, we obtain the form:

$$\rho(\vartheta) = \frac{1}{\sqrt{1 - (\sin\vartheta)^2}} \tag{7.1}$$

with $0 \leq (\sin\vartheta)^2 \leq 1$, like ε^2 and $\frac{v^2}{c^2}$.

Alternatively, we can use $\cos^2\vartheta = 1 - \sin^2\vartheta$ which then yields $\rho(\vartheta) = \frac{1}{\sqrt{1-2(\sin\vartheta)^2}}$. This has a nice geometrical meaning (Fig. 7.2). Let two lines be drawn in a rectangular coordinate system through the origin, one in the first octant and its

Fig. 7.2 Application of areas to halfchords

$$\rho(\vartheta) = \frac{1}{\sqrt{\blacksquare - 2\blacksquare}}$$

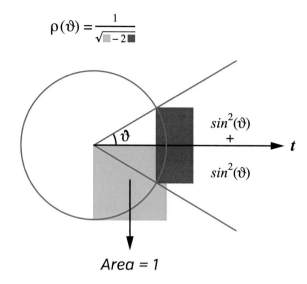

$sin^2(\vartheta) + sin^2(\vartheta)$

Area = 1

mirror image in the last octant over the horizontal axes, and a unit circle. Next, a line is drawn connecting the points of intersection of the lines and the unit circle called the chord. Since the sine function is defined as half the chord, $\sin^2\vartheta$ corresponds to the square applied to the sine (in the sense of application of areas of the Pythagoreans). Then $2\sin^2\vartheta$ corresponds to the two red squares applied to each half chord. For $0°$ yields $\rho(\vartheta) = \frac{1}{\sqrt{1-2(\sin\vartheta)^2}} = 1$ and for $45°$ $\rho(\vartheta) \to \frac{1}{0}$. This corresponds, respectively, to $v = 0$ and $v = c$ in Lorentz transformations (or $\varepsilon = 0$ and $\varepsilon = 1$ in the form of Henri Poincaré). If the green square is rotated by $45°$ in Fig. 7.2 one has a direct visual proof of this fact. In this way we convert circular functions and motion into rectilinear motion (and back) as the Greek knew. Starting from the hyperbola this may well be the shortest introduction to SRT. Once we understand our semi-rigid rules it becomes crystal clear.

In analogy with Eq. 5.9 for $B = \frac{1}{\sqrt{2}}$ or $\frac{\sqrt{2}}{2}$ we obtain $\rho(\vartheta) = \frac{1}{\sqrt{1-\frac{1}{B^2}(\sin\vartheta)^2}} = \frac{1}{\sqrt{1-2(\sin\vartheta)^2}}$. The side of the green square is 1 and its diagonal is $\sqrt{2}$. The side of the red square at $45°$ is $\frac{\sqrt{2}}{2}$ and its diagonal is 1. The green square is then equal to four halves of the red ones (cut along their diagonal). From the point of view of application of areas, this is a precise fit, in Greek: παραβολη or parabola.

It is noted that in the early years of the Lorentz transformation, actually until Einstein's SRT, the transformation did not have a physical significance, but was used as mathematical aid related to measurable facts in nature. Equations 5.8 and 5. 9 on the other hand have been developed from the start as a description and natural measure for natural shapes. But they are structurally the same.

A Theorem of Gauss and a Remarkable Year

In the form, $\rho(\vartheta) = \dfrac{1}{\sqrt{1 - \frac{1}{B^2}(\sin \vartheta)^2}}$ the coefficient $\frac{1}{B}$ takes the role of k in elliptic integrals' integrands for $B \neq 0$. Elliptic functions are based on ellipses, and are in this sense a generalization of classical trigonometric functions. In their well-known form the integrand of elliptic integrals of the first kind is given by $\dfrac{1}{\sqrt{1 - k^2 \sin^2 \vartheta}}$, and those of the second kind by $\sqrt{1 - k^2 \sin^2 \vartheta}$.

Gauss came extremely close to Eqs. 5.1 and 5.8 in his studies on the Arithmetic-Geometric Mean *AGM* of two numbers m and n around 1800. The arithmetic-geometric mean *AGM* (m,n) or *M* (m,n) is an iterative method, where the arithmetic mean *AM* and geometric mean *GM* are computed and then used as input to the next iteration. Gauss used this method for fast computations. For example, $M(\sqrt{2}, 1)$ can be computed up to 19 (!) decimal places in only four steps. From Gauss [160]:

"Let m and n be two positive magnitudes, and let us set:

$$m' = \frac{1}{2}(m + n), \quad n' = \sqrt{mn}$$

so that m' and n' are, respectively, the arithmetic and geometric mean between m and n. We assume that the geometric mean is always taken as positive. Further, let:

$$m'' = \frac{1}{2}(m' + n'), \quad n'' = \sqrt{m'n'}$$

$$m''' = \frac{1}{2}(m'' + n''), \quad n''' = \sqrt{m''n''},$$

and so forth. The series m, m', m'', m''', etc. and n, n', n'', n'' etc. tend towards a common limit, which we will denote by M and which will we will call simply, the arithmetic-geometric mean between m and n. We will show that $\frac{1}{M}$ is the value of the integral,

$$\int \frac{dT}{2\pi\sqrt{mm \, \cos T^2 + nn \, \sin T^2}} \tag{7.2}$$

taken from $T = 0$ to T $= 360°$."

In 1818 Gauss applied the Arithmetic-Geometric Mean to compute the complete elliptic integral, with the following relation:

$$\frac{1}{\pi} \int_0^\pi \frac{d\vartheta}{\sqrt{1 - x^2 \cos^2 \vartheta}} = \frac{1}{M(1, \sqrt{1 - x^2})} = \frac{1}{M(1 + x; 1 - x)} \tag{7.3}$$

This integral is encountered in determining secular perturbation by one planet onto the other planet:

$$\frac{1}{2\pi} \int_0^{2\pi} \frac{dT}{\sqrt{m^2 \cos^2 T + n^2 \sin^2 T}} = \frac{1}{M(m,n)} \tag{7.4}$$

Gauss used the same methods for computing elliptic integrals of the second kind:

$$\frac{1}{2\pi} \int_0^{2\pi} \frac{\cos^2 T - \sin^2 T}{\sqrt{m^2 \cos^2 T + n^2 \sin^2 T}} dT \tag{7.5}$$

This resonates with Lamé curves in polar representation. Remarkably the paper of Gauss and Lamé's book *Examen* were published in the same year 1818 and the combination of both ideas could have led naturally to the Superformula. Indeed, if Gauss had developed his transformations in a slightly different direction, using general exponents like Lamé did, he would have arrived naturally at superellipses (since the integrand of the Gauss form is an ellipse). Adding $m/4$, Gauss would have arrived at the Superformula. If Lamé would have used Gauss' polar version...

FRLM Geometry and FRLW Metrics

Gauss was very close to Gielis transformations in his study of the *AGM*, and so was Gabriel Lamé (and even René Descartes with the Folium). For Lamé curves the Euclidean metric is obtained for $n = 2$, whereby the length of a line is expressed using Pythagoras' theorem or $\sqrt{x^2 + y^2}$, or $ds^2 = \sqrt{dx^2 + dy^2}$ in differential form. In his historical Habilitationsschrift Bernhard Riemann (1828–1868) generalized Gauss' work for surfaces to higher dimensional manifolds. At a certain point in the text he mentions the possibility to use a fourth power instead of a square, to measure distances on manifolds between two infinitely close points,

Für den Raum wird, wenn man die Lage der Punkte durch rechtwinklige Coordinaten ausdrückt, $ds = \sqrt{\sum (dx)^2}$; der Raum ist also unter diesem einfachsten Falle enthalten. Der nächst einfache Fall würde wohl die Mannigfaltigkeiten umfassen, in welchen sich das Linienelement durch die vierte Wurzel aus einem Differentialausdrucke vierten Grades ausdrücken lässt. Die Untersuchung dieser allgemeinern Gattung würde zwar keine wesentlich andere Principien erfordern, aber ziemlich zeitraubend sein und verhältnissmässig auf die Lehre vom Raume wenig neues Licht werfen, zumal da sich die Resultate nicht geometrisch ausdrücken lassen; ich beschränke mich daher auf die Mannigfaltigkeiten, wo das Linienelement durch die Quadratwurzel aus einem Differentialausdruck zweiten Grades ausgedrückt wird [161].

Riemann's view was studied fifty years later by Paul Finsler [162] which led to the development of Riemann-Finsler geometry with norms based on powers other than two (as in the case of Riemannian geometry, or the Pythagorean measure).

Finsler manifolds are those for which the metric structure is given by a collection of convex symmetric bodies in the various tangent spaces. For the Riemannian case it is a (hyper) ellipsoid. There are already many applications of Finsler geometry in physics and biology [163].

The simplest Riemann-Finsler metrics, for any dimension n and any power p, are defined as [12; 100]:

$$\text{Riemann-Finsler metric} \quad ds = \left\{ \sum_{i=1}^{n} (dx^i)^p \right\}^{\frac{1}{p}} \tag{7.12}$$

And for $p = 2$

$$\text{Euclidean metric } ds = \left\{ \sum_{i=1}^{n} (dx^i)^2 \right\}^{\frac{1}{2}} \tag{7.13}$$

Minkowski created the formal tools to study problems about convex regions and bodies. This led to Minkowski geometry, which is *"the kind of geometry in which lengths are measured differently in different directions... Unit circles and spheres are not the familiar round objects from Euclidean geometry, but are some other convex shape, called the* unit *ball"* [28]. Cityblock L_1, the Euclidean metric L_2, and the max norm L_∞ are well-known examples of pth root metric of Minkowski where the unit circles are special cases of Lamé's supercircles. A.C. Thompson discriminates between *Minkowski* geometry and *Minkowskian* geometry, the latter related to SRT. They can now be united. It would in my view be reasonable to refer to this geometry, whatever its use or specification (bodies, submanifolds, tangent spaces....) as Finsler-Riemann-Lamé-Minkowski or *FRLM* geometry.

It was pointed out that Gielis transformations on functions, e.g. on the circle and sphere, are quite alike the transformation on the Pythagorean metric to go from Euclidean space to metrics for Riemannian spaces of constant curvature. An 'isotropic' manifold or a space of constant curvature amounts to the axiom of free mobility, whereby the lengths of rigid measuring rods are independent of their locations in space. Riemann defined such spaces using the metric:

$$ds = \frac{1}{1 + \frac{k}{4}\sum x^2} \sqrt{\sum dx^2} \quad or: \quad ds^2 = \frac{(dx^1)^2 + (dx^2)^2 + \cdots + (dx^n)^2}{\left\{ 1 + \frac{k}{4} \left[(x^1)^2 + (x^2)^2 + \cdots (x^n)^2 \right] \right\}^2} \tag{7.14}$$

The Pythagorean theorem defines the best-known example of a manifold of constant curvature, namely circles and spheres. We now transform these circles and spheres into a variety of spaces with a global anisotropy. However, the notion of anisotropy (vis-à-vis isotropy) then needs rethinking because all transformed circles (or n-spheres) are unit circles (or unit n-spheres) for certain classes of anisotropic spaces.

One application of constant curvature surfaces is space-time models of the Big Bang type, equipped with Friedman-Lemaître-Robertson-Walker FLRW metrics given by:

$$ds^2 = -dt^2 + \frac{1}{\left[c(t).\left\{1 + \frac{k}{4}[x^2 + y^2 + z^2]\right\}\right]^2}\left(dx^2 + dy^2 + dz^2\right) \qquad (7.15)$$

with $k = 0$, 1 or -1 for the Euclidean, spherical/elliptical and hyperbolic case, respectively, with a Pythagorean-compact metric.

Representing Dimensions

These metrics are given in $\mathbb{R}^4$ with local coordinates (x, y, z, t) whereby the space-slices of the full space-time at any given time t are Riemannian 3D-spaces of constant curvature $c(t)$. By way of visualization in 2D rather than in 4D representation, such metrics are carried for instance by surfaces of revolution in 3D-space. In space-time metrics like FLRW one may observe, on the one hand, the conformal deformations of flat space-metrics to metrics of constant spatial curvature ($=; <$ or > 0) and a hyperbolic twist in the line-direction, which might include a speed of light-normalization, which all in all amount to formal deformations of the theorem of Pythagoras quite alike the Gielis transformations relate to the equation of Euclidean circles or spheres [12; 100].

It is noted that Gielis transformations also redefine rotations and revolutions, where paths swept by certain curves or surfaces can take any form as well and one might easily imagine the extension to Riemann-Finsler metrics. The parametric equation for 3D surfaces is known as a "spherical product" of two perpendicular shapes. Originally this was proposed for superquadrics, but is now generalized for supershapes, generalizing spherical coordinates in $\mathbb{E}^3$ and in $\mathbb{E}^4$ [164], and twisted surfaces in $\mathbb{E}^3$ and $\mathbb{E}^3_1$ [120; 121].

With respect to four-dimensional space, this parametric description of 3D shapes is based on two perpendicular sections (in the simplest case). A pyramid for example has in one direction a triangle, in the perpendicular direction a square as cross section (or a circle in the case of a cone). With both perpendicular 2D cross sections based on our transformations, this provides two times two different possible measures or a total of four different directions. The planes can be real, complex, or the combination thereof ($\mathbb{R}^2 \times \mathbb{R}^2$, $\mathbb{R}^2 \times \mathbb{C}$ or $\mathbb{C} \times \mathbb{C}$).

Only in the case of isotropic spaces, do (at least) two of these coincide, reducing to our classical Cartesian or spherical coordinate system. The perpendicular planes allow for different ways of measuring in four different directions, which may be space or time, and which can vary in the course of movement or expansion in one or more directions, with some form of normalization as needed.

Transformations in Euclidean Geometry

There are fundamentally three ways in which these transformations can be studied or used. The first two relate to the *FRLM*. The Lamé-Minkowski *LM* part relates to the study of such shapes/transformations to define a geometry, like Minkowski geometry which provide a fascinating study in their own right. These shapes can be used to study the intrinsic and extrinsic geometry of manifolds, as convex (or concave) bodies in the tangent spaces or bundles, the Riemann-Finsler *FR* part.

Traditionally surfaces are treated as manifolds, studying the intrinsic properties like Gaussian curvature. In the past 50 years, the field of submanifolds has developed rapidly, whereby manifolds can be considered as embedded in other manifolds. The simplest example is to view manifolds in ambient Euclidean spaces. An embryo inside an embryo sac can also be studied as submanifolds in manifolds, which can be continued in any direction. Curvatures of the extrinsic geometry of submanifolds include the mean curvature $H = AM = (k_1 + k_2)/2$ and Casorati curvature $C = (k_1^2 + k_2^2)/2$. These are related to the form or shape of the submanifolds in their ambients spaces and the inequalities between curvatures of intrinsic and extrinsic nature determine the natural curvature conditions. Our transformation describes both the coordinate systems of the object, as well as the space in which it lives, for example in Euclidean space, or in a sphere.

The third way is then indeed simply to consider our transformations in the framework of Euclidean geometry and space, which immediately links geometry with the natural sciences. The shapes can then be considered as embedded in a Euclidean space [165]. L. Verstraelen states that: "*By application of the corresponding Gielis transformations to the "most natural" curves and surfaces of Euclidean geometry (e.g., for dimension 2: the circles amongst the closed curves and the logarithmic spirals amongst the non-closed curves), do result many of the forms that we do observe in nature -in biology, cristallography, physics, chemistry, etc.*" [42].

The three ways are closely linked and in this sense Gielis may be considered as an acronym for *Generalized Intrinsic and Extrinsic Lengths In Submanifolds*. Generalized or Geometric, and Length refers to anything measurable. Nomen est Omen.

In Chap. 3 it was shown how central the unit element is in Euclidean geometry, for understanding differentiation, conic sections, series and special functions. In nature this unit may differ depending on the direction. A specific transformation (Eq. 5.8) is a generalization of Lamé curves and of the Pythagorean Theorem, and if we would broaden the scope of Euclid's definitions to include "our" unit circles (which can be constructed using an elastic (or a spring) stretched according to Eqs.

5.8 and 5.9 to draw any "unit" circle), many of the classic results in geometry and extensions, might be obtained in a (relatively) simple way.

This is pretty similar to the situation one century ago: *"What, then, has changed with the work of Lorentz? Poincaré's answer was that there are now two principles that could serve to define space: the old one involving rigid bodies, and a new one to do with the transformations that do not alter our differential equations. They are not essentially different, because both are statements about objects around us, but the new one is an experimental truth. To make geometry immune to revision by experimenters, physical relativity must become a convention concerning distant objects. Then, whereas our conventional knowledge of geometry was formerly rooted in the group of Euclidean isometries, it could now be the Lorentz group that preserves our equations, at the price of placing us in a four-dimensional space"* [158].

It provides a dual view on rigid and semi-rigid, both valid. Natural organisms and shapes can be studied in their own geometry, and with immediate representation in ours with circles and circular functions. With different unit circles we obtain different unit elements.

Unveiling the Secrets of Grandi Curves

Lamé-curves can describe natural shapes with square or rectangular symmetry. For different (integer or non-integer) symmetries the Gielis Formula was developed and transcendental functions were needed, since it was not possible to generalize Lamé curves directly to any symmetry. Ultimately however, we would like to develop a simple rule of going from Cartesian to polar coordinates and back (much in the same way as the Folium of Descartes), without loss of generality for our shapes. Would this be possible in any way? The answer is yes, and once more the secret lies in trigonometric functions and their relation to description of flowers.

The oldest and most useful mathematical representation of flowers are Rhodonea curves, discovered by Guido Grandi and communicated in a letter to Leibniz in 1713, three centuries ago [166]. In the Philosophical Transactions of the Royal Society, Guido Grandi's *A collection of Geometrical Flowers* was published, on January 1, 1722 [167]. Ultimately this resulted in Guido Grandi's book *Flores geometrici ex Rhodonearum, et Cloeliarum curvarum descriptione resultantes* of 1728 [168]. Clelia curves are the 3D analogues of Rhodonea curves (Fig. 7.3).

The observation that in Rhodonea or Grandi curves $\rho = \cos m\vartheta$ or $\rho = \sin m\vartheta$ the arguments of the angle specify the frequency or number of petals was applied to the polar representation of the Lamé curves or superellipses, which led to the Superformula or Gielis Formula. The pivotal step to the superformula was rewriting Lamé curves in polar coordinates and generalizing the symmetry from 4 to any number, as in Grandi curves. The parameter $\frac{m}{4}$ aligns with the symmetries of the

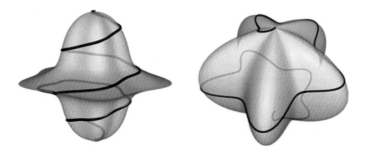

Fig. 7.3 Clelia curves developed onto supershapes. Copyright Wendy Goemans

shapes; pentagons, and starfish are defined with $m = 5$, so that the orthogonal coordinate system with right angles 90° is folded into five sectors of 72° each. When $m = \frac{5}{2}$ the spacing between the sectors is 144° and the shape closes in two rotations or 720°.

The superformula in its original form makes use of transcendental functions. What hitherto was missing is the reverse step, namely to express such shapes in Cartesian coordinates, to convert the transcendental functions into rational functions. This can be achieved using Chebyshev polynomials introduced by Pafnuty Chebyshev in the mid 19th century [169; 170]. We mentioned these curves already a few times. There is a close relationship between Chebyshev polynomials and many classic trigonometric identities.

Using $x = \cos\vartheta$, they allow the study of polynomials with transcendental functions as algebraic functions in one variable.

Chebyshev polynomials of first kind T and second kind U of order n are defined as:

$$T_m(\cos\vartheta) = \cos(m\vartheta) \tag{7.16}$$

$$U_m(\cos\vartheta) = \frac{\sin[(m+1)\vartheta]}{\sin\vartheta} \tag{7.17}$$

with:

$$T_0(x) = 1; T_1(x) = x, T_2(x) = x^2 - 1; T_3(x) = 4x^3 - 3x, \ldots,$$

This leads to rewriting Eq. 5.8 as follows:

$$\varrho(x) = \frac{1}{\sqrt[n_1]{|T_m(x)|^{n_2} + \left|\sqrt{1 - x^2}U_{m-1}(x)\right|^{n_3}}} \tag{7.18}$$

In this way it can be shown that for a suitable choice of parameters (exponents n and symmetry parameter m are integers) the superformula is an algebraic expression avoiding transcendent functions. It aligns perfectly with our goal to define compact, yet precise descriptions. The word precise is what it is meant to mean: precise and accurate. This inverse step is based on the simple observation that Grandi curves $\rho = \cos m\vartheta$ ARE Chebyshev functions of the first kind $T_m(x) = \cos m\vartheta$.

If we wish to frame it into taxonomical language: *"Research has shown that* Rhodonea cleliae *and* Polynomia oligomiae, *considered as very different species since they were first described, not only belong to the same genus* (Grandia), *but they are the same species and even the same cultivar"*. A letter of 300 years ago, in which Guido Grandi communicated to Leibniz the discovery of Rhodonea curves, inspired both the development of the superformula in polar coordinates and generalized Pythagorean Theorem in 1997, as well as the inverse transformation into Chebyshev polynomials two decades later, closing the loop. From this we can build k-type curves and polynomials (oligomials) of order k, without having to resort to infinite series.

Orthogonial Polynomials and Phyllotaxy in Plants

With further generalizations (for example multivariate and multi-argument polynomials), a wide variety of natural and abstract shapes can now be considered as algebraic functions (for appropriate choice of parameters) with a one-on-one transformation polar-Cartesian. Once more, all the ingredients were available in the 19th century, a time when mathematics had a very close interaction with practice. Orthogonal polynomials were developed by Chebyshev from 1854 onwards, based on practical considerations of mappings and best-possible approximations of curves. The success was also based on the close interaction with his friend and colleague Charles Hermite, and his student Markov.

This French connection was no coincidence because of the close connection between Paris and Saint-Petersburg and the emphasis on combining theory and practice. Gabriel Lamé had been a professor in Saint-Petersburg from 1820–1830 [171]. When considering trigonometric functions and Chebyshev polynomials, there is a direct connection to cosine Fourier series (or sine Fourier series), and thus to the many applications in modern technology, such as .jpeg format for compressing images, and the .mp3 format for compressing music.

Returning to botany and phyllotaxy: there is a complete and direct correspondence between our curves and Fibonacci and Lucas numbers and polynomials. There is a direct relation between Chebyshev polynomials and Lucas L_n and Fibonacci numbers F_n and their wide use in plant phyllotaxy. They can all be considered as special cases of the homogeneous linear second order difference equation with constant coefficients $u_0; u_1; u_{n+1} = au_n + bu_{n-1}$, for $n \leq 1$ [172]. Fibonacci numbers F_n arise for $a = b = 1; u_0 = 0; u_1 = 1$. For $a = b = 1; u_0 = 2; u_1 = 1$, we obtain Lucas numbers L_n. If a and b are polynomials in x, a sequence of polynomials is generated. In particular if $a = 2x$ and $b = -1$, we obtain Chebyshev polynomials. They are of the first kind $T_n(x)$ for $u_0 = 1; u_1 = x$, and of the second kind $U_n(x)$ for $u_0 = 1; u_1 = 2x$.

Therefore, if in Chebyshev polynomials $i = \sqrt{-1}$ is used with $x = \frac{i}{2}$ the results are Lucas numbers L_n for Chebyshev polynomials of the first kind T_n, and Fibonacci numbers F_n for those of the second kind U_n.

$$L_n = 2i^{-n}T_n\left(\frac{i}{2}\right) \qquad F_{n+1} = i^{-n}U_n\left(\frac{i}{2}\right) \tag{7.19}$$

This also clarifies the direct relation between Fibonacci and Lucas numbers, with the geometry of phyllotaxy given by Eq. 5.8 with m a rational number ($m = 5/2$, 8/3, for example... F_{n+2}/F_n). These give the number of angles in the numerator and the number of rotations needed to close, as the inverse of the generally used F_n/F_{n+2} in phyllotaxy. As a remark, instead of starting from a finite difference equation, one can generalize complex numbers and derive Chebyshev polynomials in this way.

Grandi and Gielis curves take the study of plants and natural organisms further into the realms of the three pillars of mathematics, namely geometry, algebra and analysis. Our goal of developing compact representations of natural shapes is achieved, either using the Superformula or the inverse transformation via Chebyshev polynomials. We now can express a wide variety of natural and abstract shapes using monomials, which may be of different nature. Furthermore, the analytic expression immediately suggests a generalization in two ways: First the classical transformation $x = \cos \vartheta$ can be generalized with $x = \cos f(\vartheta)$, where $f(\vartheta)$ can involve a time component. Second, using the coordinate functions of supershapes,

$$\varrho(\vartheta) = \frac{\cos m\vartheta}{\sqrt[n_1]{\left|\frac{1}{A}\cos\left(\frac{m}{4}\vartheta\right)\right|^{n_2} \pm \left|\frac{1}{B}\sin\left(\frac{m}{4}\vartheta\right)\right|^{n_3}}} \tag{7.20}$$

$$\varrho(\vartheta) = \frac{\sin m\vartheta}{\sqrt[n_1]{\left|\frac{1}{A}\cos\left(\frac{m}{4}\vartheta\right)\right|^{n_2} \pm \left|\frac{1}{B}\sin\left(\frac{m}{4}\vartheta\right)\right|^{n_3}}} \tag{7.21}$$

we can immediately generalize Chebyshev polynomials as:

$$\varrho(\arccos(x)) = \frac{T_m(x)}{\sqrt[n_1]{|T_m(x)|^{n_2} + \left|\sqrt{1-x^2}U_{m-1}(x)\right|^{n_3}}} \tag{7.22}$$

Instead of $f(\vartheta) = R$ we may indeed use Chebyshev polynomials, and in this way this family of polynomials is generalized. By using substitutions of the type $x = \cos(f(\vartheta))$, for reasonable choices of $f(\vartheta)$ our generalized form of the Gielis Formula works with Chebyshev as well.

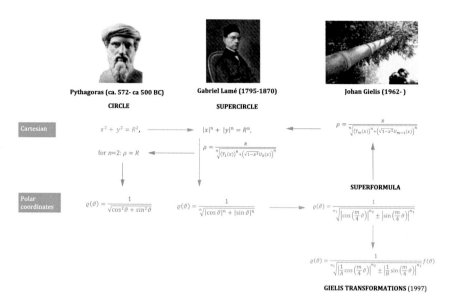

Fig. 7.4 From Pythagoras to Gielis transformations and back

In this way we retain the Pythagorean-compact expression, but we can use transcendental functions or rational ones. In this way we close the scheme, started by generalizing Lamé in polar coordinates, and finalized by rewriting it in Chebyshev polynomials (Fig. 7.4). For a suitable choice of parameters, the shapes defined by Eq. 5.8 can then be treated as rational functions.

Grandi-curves provide the common descent for both a uniform description of natural shapes in a very broad sense, and for various mathematical-geometrical methods. For this class of shapes we have the choice of three expressions. First, in a Pythagorean structure, second as an infinite series, via e.g. Fourier series or Taylor series expansion or similar, and third as a finite polynomial. This means that we can switch between Cartesian coordinates and transcendental trigonometric functions, with many applications in science and technology.

Part IV
Κατασκευή—Constructio

Chapter 8
(Meta)Harmony

En résumé, lorsqu'une classe de phénomènes physiques dépend des variations d'un certaine fonction-de-point, c'est presque uniquement par son paramètre différentiel du second ordre que cette fonction intervient. Comme si ce paramètre était une dérivée naturelle, plus essentielle, plus simple, et en même temps plus complète, que toutes les dérivées partielles, choisis plus ou moins arbitrairement, que'on a l'habitude de considérer.

Gabriel Lamé (1836)

Harmonic and Metaharmonic Solutions

All plant-shapes generated by one deterministic L-system are identical, so stochastic L-systems were introduced in order to produce more variation. The drawback is that, when a desired plant shape is given, one cannot find, in an algorithmic or otherwise mathematically precise way, the L-system that will generate it [94]. This is true for fractals, supershapes, R-functions or any other model (and in all of science). A plant however, is more than meets the eye. It is a complex organism, the result of the precise interplay between geometry, genes and enzymes, and many other substances and metabolites, growing in specific climatic and edaphic conditions, subject to many biotic and abiotic factors, prone to diseases and in need of pollinators, with limited or abundant availability of water, light and nutrients. Its genetic constitution is the result of million years of evolution at work and has incorporated all mechanisms necessary to develop and grow successfully.

Hence, our model needs to be much more than a virtual plant that looks nice (see Fig. 6.8). The model we need is that of an "abstract plant", not a virtual plant, where certain model-parts are controlled by suitable parameters. Plants are continuously communicating with an ever-changing environment where macro and micro-effects interact and the parameters in the abstract plant enable us to measure the effect of these interactions, hence introducing parameters and choosing values of them with the aim of obtaining a virtual plant resembling a given specimen contradicts our

© Atlantis Press and the author(s) 2017
J. Gielis, *The Geometrical Beauty of Plants*,
DOI 10.2991/978-94-6239-151-2_8

aim. Our "abstract" plant in the study of the morphogenesis is a time sequence of adaptations to internal and external stresses acting on the plants. The model constructed should be generic, i.e. free of in-built constraints depending on results or theories contained in those to be tested in the actual research [94]. An abstract tree will be different from an abstract cactus, but the mathematical model underlying both is the same.

One of the advantages of our model is a continuous transformation. A plant is a combination of transitional forms: cotyledons, vegetative leaves and their polymorphism, sepals, petals, stamen and pistils are all variations and transitions of leaves. This resonates well with Goethe's studies of plants: *Alles ist Blatt, und durch diese Einfachheit wird die grösste Mannigfaltigkeit möglich.* "Everything is Leaf, and from this simple fact, the greatest diversity becomes possible".

There is a nice parallelism with geometry. Riemann's manifolds or Mannigfaltigkeiten are described via their intrinsic characteristics, their DNA if you will. The actual shape the manifold will assume in some space, for example a physical space, depends on the close and continuous interaction with the environment in the widest sense, and then the actual shape will depend upon the extrinsic characteristics. Actual shapes are a result of the interplay between the intrinsic and extrinsic characteristics, striving towards equilibrium, static or dynamic. The way soap bubbles and soap films form and stabilize as a dynamical process, is the result of such interplay. Because of the isotropic nature of soap the molecules are pretty free to organize themselves minimizing stress along the surface, or at least distribute the stress as evenly as possible. The shape of a soap bubble and soap film is in this sense very similar to chords of a guitar, or a drum: once set into motion, the most natural tendency is to go back to static equilibrium. The study of minimal surfaces and dynamic equilibria are governed by differential equations, the heart of physics.

Again, Gabriel Lamé was at the forefront of these developments, following in the footsteps of among others Euler and Laplace. In his book *Coordonnées Curvilignes* he introduced the differential parameters of the first Δ_1 (or ∇) and second kind Δ_2 (or Δ). The latter is the Laplacian, and it is central in potential theory, heat distribution, plate vibrations, the wave equation and so on. It is directly related to the mean curvature H on the surface. The Laplacian is a crucial ingredient of many partial differential equations, such as the equations of Laplace, Poisson and Helmholtz, the wave and Schrödinger equations. In polar coordinates:

$$\Delta u = \frac{\partial^2 u}{\partial \rho^2} + \frac{1}{\rho}\frac{\partial u}{\partial \varrho} + \frac{1}{\rho^2}\frac{\partial^2 u}{\partial \vartheta^2}$$

In the words (and notation) of Gabriel Lamé:

In the theory of the potential P, the general equation is:

$$\Delta_2 P = 0$$

In the analytic theory of heat, the temperature V is determined by the equations:

$$\Delta_2 V = 0 \quad \text{in steady state}$$
$$\Delta_2 V = k\frac{dV}{dt} \quad \text{in cooling}$$

In the mathematical theory of elasticity of homogeneous solids, the cubic dilatation θ is determined by the equations:

$$\Delta_2 \theta = 0 \quad \text{in steady state}$$
$$\frac{d^2\theta}{dt^2} = a^2 \Delta_2 \theta \quad \text{under vibrations}$$

The proper selection of the displacement of molecules and the components of elastic forces in steady state all follow the equations

$$\Delta_2 \Delta_2 F = 0$$

In summary, when a class of phenomena in physics depends on variations of a certain point-to-point function, it is almost always through its second order differential parameter that this function operates, as if this parameter is the most natural, most essential, simplest and at the same time the most complete, of all partial derivatives, that one has the habit to consider, but the choice is more or less arbitrary.

Lamé developed deep insights in mathematical physics with far-reaching consequences up to the present day. Lamé envisaged that, from a mathematical point of view, the study of a physical system amounts to the study of a system of curvilinear coordinates, adapted to the given physical situation. The mathematical world of curvilinear coordinates is then like a model of the world of physical systems. René Guitart remarks: *"The importance of the general idea of considering all curvilinear coordinate systems, and not only the four or five known (Cartesian, oblique, cylindrical, spherical, bipolar) around 1830 is really considerable. One may compare this with Descartes, who instead of only 10 or 12 curves, known in Ancient times, passed over into a real infinitude of curves meant to represent physical systems"* [85].

Gaston Darboux wrote, in his overview of 19th century geometry: *"At this moment, one of the most penetrating of geometers, according to the judgement of Jacobi, Gabriel Lamé, who, like Charles Sturm had commenced with pure geometry and ... rising to the idea of curvilinear coordinates in spaces, became the creator of a wholly new theory destined to receive in mathematical physics the most varied applications"* [173].

The study of a physical problem, adapted with the appropriate system of curvilinear coordinates, then becomes the characterization of the system of differential invariants, which equals the calculation of the Laplacian in curvilinear coordinates. In Lamé's view only one equation then needs to be calculated, namely the Poisson equation in curvilinear coordinates with boundary conditions. Other equations and laws are reduced to special cases. It is very relevant for botany as well. Paul B. Green envisioned combining calculus and development in plants and plant organs, in which solutions to the differential equations generated local deviations resulting in organs, e.g. stamens or petals in a ring or whorl [174].

Gielis curves and surfaces carry natural curvilinear coordinate systems adapted to the system under study, reminiscent of Gabriel Lamé's Unique Rational Science. Lamé's Science Rationelle Unique, Universal Natural Shapes, and a geometrical theory of morphogenesis all resonate along the same lines of providing a geometrical picture of the world. Universal Natural Shapes give a unifying and coherent geometrical way of measuring and modeling shapes in line with the views of Lamé, Riemann and Helmholtz on science.

The Mathematical Physics Project of Fourier

The memoir of Jean-Baptiste Fourier aimed at providing a solution to a specific problem in mathematical physics, the heat equation, using circular trigonometric functions [175]. René Thom wrote that the applications of mathematics in the science in almost all cases use pre-existing mathematical results. The work of Fourier was for him the only counterexample in science. It is rooted in physics and Fourier developed his own, original methods using trigonometric functions based on the circle. To put this in perspective: Newton had succeeded in determining the curvature of any planar curve, using osculating circles in any point, and the radius of these circles became the measure of curvature. This then, could be linked directly, up to a constant, to mechanical forces in nature. In one stroke, Fourier provided a global determination of curves and functions, using a sum of a circle and circular functions (Fourier series). It was groundbreaking work, to say the least, generalizing the solution of vibrating strings (1D in a 2D domain) in terms of harmonics, for planar domains in 3D Euclidean spaces.

Despite the widespread use of Fourier series in science it remained an open problem whether the idea of Fourier to approximate functions with any degree of precision using Fourier series was valid. It was only in 1966 that Lennart Carleson could prove the point-wise convergence of Fourier series, almost everywhere, except in points of measure zero.

This settled a long-standing problem, but it did not complete Fourier's project on mathematical physics. The original *problem* was a quadruple of a boundary value problem BVP, boundary conditions BC (Dirichlet, Neumann, Robin), a given function on the boundary BF, and a domain *D,* and the *solution* in terms of Fourier series. Since Fourier series make use of trigonometric functions, the solution was found on a circular domain, but for the next two centuries analytical solutions remained possible only for a few domains, since one lacked a general method to reduce regular and irregular shapes to circles or circular functions. A variety of methods were developed, very few analytic, most numerical, but none of these methods could use the classical Fourier projection method.

Solving BVP with Fourier Series

It was only in 2007, exactly 200 years after Fourier's original memoir that it was shown that the solution of Boundary Value Problems for any normal polar domain could be expressed in Fourier series. P.E. Ricci, P. Natalini and D. Caratelli generalized the Laplacian, using the Gielis transformation to transform any domain into a circle, and obtained Fourier solutions for a variety of canonical problems in mathematical physics [176, 177]. The domains not only included normal polar planar domains, but also 3D domains and Riemann surfaces. Later these solutions were extended to annular domains and shells.

Since the original paper in 2007, solutions using only 19th century mathematics [Fourier series, Bessel (first described by Fourier) and Hankel functions] have been obtained for heat, wave, Laplace, Poisson and Helmholtz equations, under various boundary conditions on 2D and 3D domains, using the classical Fourier projection method [178–181]. A list of the most important boundary value problems (see Lamé's list):

- Wave equation $\quad a^2 \Delta v = v_{tt}$
- Heat propagation $\quad k\Delta v = v_t$
- Laplace equation $\quad \Delta v = 0$
- Poisson equation $\quad \Delta v = f$
- Helmholtz equation $\quad \Delta v + k^2 v = 0$
- Schrödinger's equation $\quad -\frac{k^2}{2m}\Delta\psi + V\psi = E\psi$
- Plate vibrations $\quad \nabla^4 + \frac{\varrho\partial^2 w}{\partial t^2} = 0$

The solutions are based on the generalization of the Laplacian for stretched coordinate systems. How this works will be illustrated for the Laplace equation for a flying bird domain [134], and for the Helmholtz equation in relation to plate vibrations.

Let us consider in the Euclidean plane the classical polar coordinate system

$$\begin{cases} x = \rho\cos\vartheta \\ y = \rho\sin\vartheta \end{cases}$$

and the starlike domain D, whose boundary ∂D is described by the equation $\rho = R(\vartheta)$, where $R(\vartheta)$ is a C^2 function in $[0, 2\pi]$, such that $m_R = min_{\vartheta \in [0,2\pi]}R(\vartheta) > 0$, and $M_R = max_{\vartheta \in [0,2\pi]}R(\vartheta) < 1$. Therefore, upon introducing the stretched radius: $r = \frac{R(\vartheta)}{\rho}$ it is straightforward to see that D satisfies $0 \leq \vartheta \leq 2\pi$, and $0 \leq r \leq 1$.

Let us consider a $C^2(D)$ function of $v(x, y) = v(\rho \cos \vartheta, \rho \sin \vartheta) = u(\rho, \vartheta)$ and the relevant Laplacian operator

$$\Delta u = \frac{\partial^2 u}{\partial \rho^2} + \frac{1}{\rho} \frac{\partial u}{\partial \varrho} + \frac{1}{\rho^2} \frac{\partial^2 u}{\partial \vartheta^2}$$

Upon setting $u(r, \vartheta) = u(rR(\vartheta), \vartheta)$, we find:

$$\frac{\partial u}{\partial \rho} = \frac{1}{R} \frac{\partial U}{\partial r}, \frac{\partial^2 u}{\partial \rho^2} = \frac{1}{R^2} \frac{\partial^2 U}{\partial r^2}, \frac{\partial u}{\partial \vartheta} = -r \frac{R'}{R} \frac{\partial U}{\partial r} + \frac{\partial U}{\partial \vartheta}$$

and

$$\frac{\partial^2 u}{\partial \vartheta^2} = \frac{R'^2 - RR''}{R^2} \frac{\partial U}{\partial r} + r^2 \frac{R'^2}{R^2} \frac{\partial^2 U}{\partial r^2} - 2r \frac{R'}{R} \frac{\partial^2 U}{\partial r \partial \vartheta} + \frac{\partial^2 U}{\partial \vartheta^2}$$

Hence, substituting, we obtain

$$\Delta u = \frac{\partial^2 u}{\partial \rho^2} + \frac{1}{\rho} \frac{\partial u}{\partial \varrho} + \frac{1}{\rho^2} \frac{\partial^2 u}{\partial \vartheta^2}$$

$$= \frac{1}{R^2} \left[1 + \frac{R'^2}{R^2} \right] \frac{\partial^2 U}{\partial r^2} + \frac{1}{rR^2} \left[1 + \frac{2R'^2 - RR''}{R^2} \right] \frac{\partial U}{\partial r} - \frac{2R'}{rR^3} \frac{\partial^2 U}{\partial r \partial \vartheta} + \frac{1}{r^2 R^2} \frac{\partial^2 U}{\partial \vartheta^2}$$

For $r = \rho, R(\vartheta) \equiv 1$ we recover the Laplacian in usual polar co-ordinates.

The Dirichlet Problem for the Laplace Equation

The first example is the Laplace equation and the interior and exterior Dirichlet problem for the Laplace equation in a starlike domain $\mathcal{D}$, whose boundary is described by the polar equation $\rho = R(\vartheta)$

$$\begin{cases} \Delta v(x, y) = 0, & (x, y) \in D, \\ v(x, y) = f(x, y), & (x, y) \in \partial D \end{cases} \qquad (**)$$

Theorem 8.1 *Let*

$$f(R(\vartheta) \cos \vartheta, R(\vartheta) \sin \vartheta) = F(\vartheta) = \sum_{m=0}^{+\infty} (a_m \cos m\vartheta + \beta_m \sin m\vartheta)$$

where

$$\left\{ \begin{array}{c} a_m \\ \beta_m \end{array} \right\} = \frac{\varepsilon_m}{2\pi} \int_0^{2\pi} F(\vartheta) \left\{ \begin{array}{c} \cos m\vartheta \\ \sin m\vartheta \end{array} \right\} d\vartheta$$

ε_m being the usual Neumann's symbol. Then, the interior boundary-value problem for the Laplace equation admits a classical solution:

$$v(x, y) \in C^2(\mathcal{D})$$

such that the following Fourier-like series expansion holds:

$$v(rR(\vartheta)\cos\vartheta, R(\vartheta)\sin\vartheta) = U(r, \vartheta) = \sum_{m=0}^{+\infty} [rR(\vartheta)]^m (A_m \cos m\vartheta + B_m \sin m\vartheta).$$

The coefficients A_m, B_m can be determined by solving the infinite linear system:

$$\sum_{m=0}^{+\infty} \begin{bmatrix} X_{n,m}^+ & Y_{n,m}^+ \\ X_{n,m}^- & Y_{n,m}^- \end{bmatrix} \cdot \begin{bmatrix} A_m \\ B_m \end{bmatrix} = \begin{bmatrix} a_n \\ \beta_n \end{bmatrix}$$

where

$$\begin{cases} X_{n,m}^{(\pm)} = \frac{\varepsilon_n}{2\pi} \int_0^{2\pi} R(\vartheta)^m \cos m\vartheta \left\{ \begin{array}{c} \cos n\vartheta \\ \sin n\vartheta \end{array} \right\} d\vartheta \\ Y_{n,m}^{(\pm)} = \frac{\varepsilon_n}{2\pi} \int_0^{2\pi} R(\vartheta)^m \sin m\vartheta \left\{ \begin{array}{c} \cos n\vartheta \\ \sin n\vartheta \end{array} \right\} d\vartheta \end{cases} \qquad (**)$$

with $m, n \in \mathbb{N}_0$

The given problem can be solved using the classic Fourier methods, with separation of variables, as shown in the proof.

Proof In the stretched co-ordinate system (r, ϑ) the domain D is transformed in the unit circle. Therefore, the usual eigenfunction method and separation of variables (with respect to ρ, ϑ) can be conveniently used. As a consequence, elementary solutions of the problems can be searched in the form:

$$u(\rho, \vartheta) = U\left(\frac{\rho}{R(\vartheta)}, \vartheta\right) = P(\rho)\Theta(\vartheta)$$

By substituting into the Laplace equation, one can find out that the functions $P(\rho)\Theta(\vartheta)$ must satisfy the ordinary differential equations:

$$\frac{d^2\Theta(\vartheta)}{d\vartheta^2} + \mu^2\Theta(\vartheta) = 0,$$

$$\rho^2 \frac{d^2P(\rho)}{d\rho^2} + \rho\frac{dP(\rho)}{d\rho} - \mu^2 P(\rho) = 0$$

respectively.

The parameter μ is a separation constant whose choice is governed by the physical requirement that at any fixed point in the plane the scalar field $u(\rho, \vartheta)$ must be single-valued. By setting $\mu = m \in \mathbb{N}_0$ we find:

$$\Theta(\vartheta) = a_m \cos m\vartheta + b_m \sin m\vartheta$$

when $a_m, b_m \in \mathbb{R}$ denote arbitrary constants. The radial function $P(.)$ satisfying the ODE can be readily expressed as follows:

$$P(\rho) = c_m \rho^m + d_m \rho^{-m} (c_m, d_m \in \mathbb{R})$$

As usual, we have to assume $d_m = 0$ for the boundedness of the solution.

Therefore, the general solution of the interior Dirichlet problem can be expressed in the form:

$$u(\rho, \vartheta) = \sum_{m=0}^{\infty} \rho^m (A_m \cos m\vartheta + B_m \sin m\vartheta)$$

As the final step, enforcing the boundary conditions yields:

$$F(\vartheta) = U(1, \vartheta) = u(R(\vartheta), \vartheta) = \sum_{m=0}^{\infty} R^m (A_m \cos m\vartheta + B_m \sin m\vartheta)$$

and finally, using Fourier' projection method, the system of linear equations easily follows ∎

Since we can define domains inside and outside by using inequalities, in a similar way the <u>exterior Dirichlet problem</u> subject to the null condition at infinity $\lim_{\rho \to \infty} v(x, y) = 0$ may be addressed

$$\begin{cases} \Delta v(x, y) = 0, & (x, y) \in \mathbb{R}^2 \backslash D, \\ v(x, y) = f(x, y), & (x, y) \in \partial D \end{cases}$$

Theorem 8.2 *Under the hypotheses of Theorem* 8.1, *the exterior boundary value problem for the Laplace equation admits a classical solution*

$$v(rR(\vartheta) \cos \vartheta, R(\vartheta) \sin \vartheta) = U(\rho^*, \vartheta) = \sum_{m=0}^{+\infty} [rR(\vartheta)]^{-m} (A_m \cos m\vartheta + B_m \sin m\vartheta).$$

The coefficients A_m, B_m can be determined by solving the infinite linear system:

$$\sum_{m=0}^{+\infty} \begin{bmatrix} X_{n,m}^+ & Y_{n,m}^+ \\ X_{n,m}^- & Y_{n,m}^- \end{bmatrix} \cdot \begin{bmatrix} A_m \\ B_m \end{bmatrix} = \begin{bmatrix} a_n \\ \beta_n \end{bmatrix},$$

where

$$
\begin{cases}
X_{n,m}^{(\pm)} = \frac{\varepsilon_n}{2\pi} \int_0^{2\pi} R(\vartheta)^{-m} \cos m\vartheta \begin{Bmatrix} \cos n\vartheta \\ \sin n\vartheta \end{Bmatrix} d\vartheta \\
Y_{n,m}^{(\pm)} = \frac{\varepsilon_n}{2\pi} \int_0^{2\pi} R(\vartheta)^{-m} \sin m\vartheta \begin{Bmatrix} \cos n\vartheta \\ \sin n\vartheta \end{Bmatrix} d\vartheta
\end{cases} \quad \text{with } m, n \in \mathbb{N}.
$$

These formulas still hold true under the assumption that the function $R(\vartheta)$ is a piecewise continuous function and the boundary data are described by square integrable functions, not necessarily continuous, so the relevant Fourier coefficients α_m, β_m are finite quantities. The methods also work for selfintersecting curves or Riemann surfaces.

Numerical Results

To assess the performance of the technique in terms of accuracy and convergence rate, the relative boundary error is evaluated using $e_N = \frac{\|U_N(1,\vartheta)-F(\vartheta)\|}{\|F(\vartheta)\|}$ with $\|\cdot\|$ denoting the usual L^2 norm, U_N the partial sum of order N relevant to the Fourier-like series expansion representing the solutions of the Dirichlet problem for the Laplace equation, and F the function describing the boundary values. The domain is that of the flying bird, a 3-type curve defined by the following parameters.

$f(\vartheta)$	$\left|\cos\left(\frac{3}{2}\right)\vartheta\right|$	$\left|\cos\left(\frac{4}{2}\right)\vartheta\right|$	$\left|\cos\left(\frac{5}{2}\right)\vartheta\right|$
n_1	2	1	1
n_2	2	1	1
n_3	2	1	1

Figure 8.1 shows the spatial distribution of the partial sum U_N of order N = 7 representing the solution of the Dirichlet problem for the Laplace equation for the domain of a flying bird and with $f(x, y) = x + \cos y$ describing the boundary data. The angular behaviour of the partial sum U_N with expansion order $N = 7$ is shown in the center and the relative boundary error e_N for N = 7 below. This is the simplest case, but in all studied BVP low orders of expansions suffice.

Flower and Plate Vibrations

The Laplace equation is the simplest of boundary value problems, describing the behaviour of a domain that is perturbed and wants to return to rest. To counteract stress on the surface, the surface reacts by breaking up the stress into harmonics to

Fig. 8.1 Spatial distribution and boundary values of the partial sum $u_{N(x,y)}$ approximating the solution of the interior Dirichlet problem for the Laplace equation in a flying bird domain. Copyright Diego Caratelli

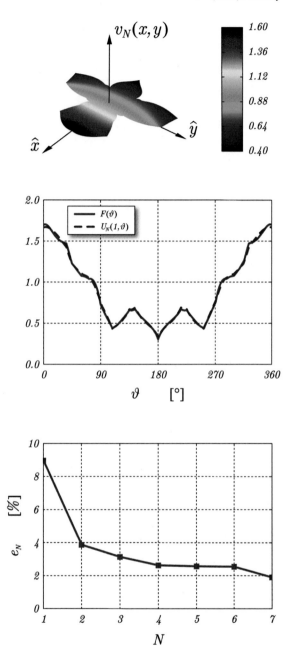

return to zero stress as soon as possible. This is a very important equation in the whole of mathematical physics, as application of the laws of Newton. For example trees and plants have trunks, branches, and leaves to deal with stress imposed by wind, rain and snow. The trunk will absorb the low frequencies, while the highest frequencies are neutralized as much as possible by fluttering leaves [182, 183]. The structure of wood inside a tree is a silent witness of the past weather conditions to which the tree had to adapt its shape continuously. Plants are very well adapted to weather conditions: the inflorescence of grasses manages to create eddy currents around its florets to catch pollen carried by the wind, as shown by Karl J. Niklas using a stroboscope, brilliant in its simplicity [184] (it was this work that made me decide to go into plant research).

In such conditions however, the stress is not a one time applied force, like plucking a string, but it is continuously or discontinuously changing over time. The next best thing for an object to deal with stress is to align it with functions that remain pretty invariant under change. The trigonometric and exponential functions are ideal for this. So finding a solution in terms of trigonometric functions or exponential functions is an important quest in mathematical physics.

Many problems of mathematical physics with partial differential equations involving time and space featuring the Laplacian, are described by the Helmholtz equation $\Delta v + k^2 v = 0$, or $\Delta v = -k^2 v$. This important equation is used to study problems in electromagnetism, acoustics and vibrations etc. Many other problems can be reduced to Helmholtz equation: for example, plate vibrations involving a fourth order Laplacian or reaction-diffusion systems, can be solved using two independent Helmholtz equations. A hyperbolic equation, like the wave equation, can be transformed into Helmholtz using the Laplace transform. Through the wave equation, there is a direct link between the Helmholtz and Schrödinger equations. Once more, we strive for an exact solution in terms of Fourier series. The "exact" solution of many classical Boundary Value Problems in terms of Fourier series can be achieved, where "exact" means that we can approximate a prescribed finite number of coefficients of the Fourier expansion of the solution as closely as we wish. This is in line with the goal that we stated: describe the shapes or phenomena in terms of conic sections (or Lamé-Gielis generalizations) and develop methods using simple, circular functions.

Fourier-Bessel Solution for the Helmholtz Equation

The displacement w of points on a vibrating plate is given by the classical differential equation

$$D\nabla^4 + \frac{\varrho\partial^2 w}{\partial t^2} = 0$$

with flexural rigidity $D = \frac{Eh^3}{12(1-v^2)}$ (E is Young's modulus, h thickness of the plate, v Poisson's ratio and ρ mass density). For free vibrations we have $w = W\cos(\omega t)$ with W a function of position coordinates. This gives: $(\nabla^4 - k^4)W = 0$, with $k^4 = \frac{\rho\omega^2}{D}$, or $(\nabla^2 + k^2).(\nabla^2 - k^2)W = 0$. The complete solution can thus be obtained by superimposing solutions to the equations [185]:

$$\begin{cases} \nabla^2 W_1 + k^2 W_1 = 0 \\ \nabla^2 W_2 - k^2 W_2 = 0 \end{cases}$$

The plate vibration problem can be solved using two Helmholtz equations. Here we present the Fourier-Bessel solution for the Helmholtz equation in our domains.

The interior Dirichlet problem for the Helmholtz equation in a starlike domain D, whose boundary is described by the polar equation $\rho = R(\vartheta)$, is:

$$\begin{cases} \Delta v(x,y) + k^2 v(x,y) = 0 & (x,y) \in D \\ v(x,y) = f(x,y) & (x,y) \in \partial D \end{cases}$$

with k the propagation constant. We then have the following theorem:

Theorem 8.3 *Let*

$$f(R(\vartheta)\cos\vartheta, R(\vartheta)\sin\vartheta) = F(\vartheta) = \sum_{m=0}^{+\infty} (a_m \cos m\vartheta + \beta_m \sin m\vartheta)$$

where

$$\begin{Bmatrix} a_m \\ \beta_m \end{Bmatrix} = \frac{\varepsilon_m}{2\pi} \int_0^{2\pi} F(\vartheta) \begin{Bmatrix} \cos m\vartheta \\ \sin m\vartheta \end{Bmatrix} d\vartheta$$

ε_m *being the usual Neumann's symbol. Then, the interior boundary-value problem for the Helmholtz equation admits a classical solution.* $v(x,y) \in C^2(D)$ *such that the following Fourier-Bessel expansion holds*

$$v(rR(\vartheta)\cos\vartheta, rR(\vartheta)\sin\vartheta) = U(r,\vartheta) = \sum_{m=0}^{+\infty} J_m(krR(\vartheta))(A_m \cos m\vartheta + B_m \sin m\vartheta)$$

with $J_m(.)$ *denoting the Bessel function of the first kind and order m. The coefficients* A_m, B_m *can be determined by solving the infinite linear system*

$$\sum_{m=0}^{+\infty} \begin{bmatrix} X_{n,m}^+ & Y_{n,m}^+ \\ X_{n,m}^- & Y_{n,m}^- \end{bmatrix} = \begin{bmatrix} a_m \\ \beta_m \end{bmatrix},$$

where

$$\begin{cases} X_{n,m}^{(\pm)} = \frac{\varepsilon_n}{2\pi} \int_0^{2\pi} J_m(kR(\vartheta)) \cos m\vartheta \begin{Bmatrix} \cos n\vartheta \\ \sin n\vartheta \end{Bmatrix} d\vartheta \\ Y_{n,m}^{(\pm)} = \frac{\varepsilon_n}{2\pi} \int_0^{2\pi} J_m(kR(\vartheta)) \sin m\vartheta \begin{Bmatrix} \cos n\vartheta \\ \sin n\vartheta \end{Bmatrix} d\vartheta \end{cases}$$

with $m, n \in \mathbb{N}_0$

Proof The proof is very similar but now will involve Bessel functions. In the stretched co-ordinate system (r, ϑ) the domain D is transformed in the unit circle. Therefore, the usual eigenfunction method and separation of variables (with respect to ρ, ϑ) can be conveniently used. As a consequence, elementary solutions of the problems can be searched in the form:

$$u(\rho, \vartheta) = U\left(\frac{\rho}{R(\vartheta)}, \vartheta\right) = P(\rho)\Theta(\vartheta)$$

By substituting into the Helmholtz equation, one can find out that the functions $P(\rho)\Theta(\vartheta)$ must satisfy the ordinary differential equations:

$$\frac{d^2\Theta(\vartheta)}{d\vartheta^2} + \mu^2\Theta(\vartheta) = 0,$$

$$\rho^2\frac{d^2P(\rho)}{d\rho^2} + \rho\frac{dP(\rho)}{d\rho} + \left[(k\rho)^2 - \mu^2\right]P(\rho) = 0$$

respectively.

The parameter μ is a separation constant whose choice is governed by the physical requirement that at any fixed point in the plane the scalar field $u(\rho, \vartheta)$ must be single-valued. By setting $\mu = m \in \mathbb{N}_0$ we find:

$$\Theta(\vartheta) = a_m \cos m\vartheta + b_m \sin m\vartheta$$

when $a_m, b_m \in \mathbb{R}$ denote arbitrary constants. The radial function $P(.)$ satisfying the ODE can be readily expressed as follows:

$$P(\rho) = c_m J_m(k\rho) + d_m Y_m(k\rho)(c_m, d_m \in \mathbb{R})$$

As usual, we have to assume $d_m = 0$ for the boundedness of the solution.

Therefore, the general solution of the interior Dirichlet problem can be expressed in the form:

$$u(\rho, \vartheta) = \sum_{m=0}^{\infty} J_m(k\rho)(A_m \cos m\vartheta + B_m \sin m\vartheta)$$

Enforcing the boundary conditions yields:

$$F(\vartheta) = U(1, \vartheta) = u(R(\vartheta), \vartheta) = \sum_{m=0}^{\infty} J_m(kR(\vartheta))(A_m \cos m\vartheta + B_m \sin m\vartheta)$$

and finally, using Fourier' projection method, the system of linear equations follows. ∎

We do have the same methods, but now Bessel functions appear, all 19th century mathematics. The developed method can be applied to the solution of differential problems in a two-dimensional flower shaped domain D, whose boundary is described by the fusion-like Gielis curve:

$$\rho = R(\vartheta) = (1-\alpha)\left|\cos\frac{m\vartheta}{2}\right| + \alpha\left[\left|\cos\left(\frac{m\vartheta}{4}\right)\right|^{n_2} \pm \left|\sin\left(\frac{m\vartheta}{4}\right)\right|^{n_3}\right]^{-\frac{1}{n_1}}$$

As an example, by using the Theorem the interior Dirichlet problem for the Helmholtz equation has been solved upon setting $m = 5$, exponent $n_i = 1$ for different values of α. Furthermore the boundary values have been assumed to be given by the radially symmetric function

$$f(x,y) = (x^2 + y^2)^\tau exp\left[(x^2 + y^2)/\sigma^2\right], \text{ with } \tau = 3 \text{ and } \sigma = 1/2$$

In particular the numerical accuracy has been tested and assessed in terms of the following boundary error.

$$e_N = \frac{\|U_N(1,\vartheta) - F(\vartheta)\|}{\|F(\vartheta)\|}$$

with $\|.\|$ denoting the usual L^2 norm, U_N the partial sum of order N relevant to the Fourier-like series expansion representing the solutions of the Dirichlet problem for the Helmholtz equation, and F the function describing the boundary values (Figs 8.2 and 8.3).

Again we see that the solution can be obtained using separation of variables and Fourier's method, with low orders of expansion.

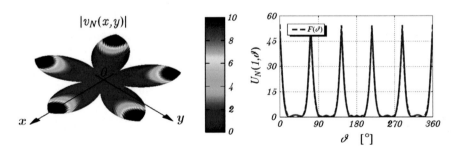

Fig. 8.2 Spatial distribution (*left*) and boundary values (*right*) of the partial sum $u_{N(x,y)}$ of order $N = 20$ approximating the solution of the interior Dirichlet problem for the Helmholtz equation in a flower shaped domains. Copyright Diego Caratelli

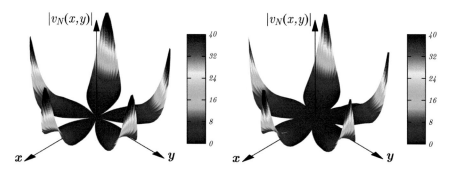

Fig. 8.3 Spatial distribution of the partial sum $u_{N(x,y)}$ of order N approximating the solution of the interior Dirichlet problem for the Helmholtz equation in the flower-shaped domain. *Left* $\alpha = 0.1$; *Right* $\alpha = 0.4$. $N = 20$ for both cases. Copyright Diego Caratelli

Metaharmony

The stretched Laplacian can be applied to any domain, in particular our Gielis domains. The solution in terms of Fourier expansions then becomes possible using separation of variables, a very classical technique, but until recently only applicable in a limited number of cases, the circle being the most illustrious example. The genius of Pythagoras was that he defined the circle in a way that directly allowed for separation of variables, since the coordinate functions, sine and cosine are orthogonal. This led to the great success of orthogonal and orthonormal functions, including Legendre functions, Fourier series, Bessel, Hankel and Hermite-Chebyshev functions. The Gielis formula retains the Pythagorean structure, so separation of variables is ensured from the start.

The Laplacian is a very important ingredient of any study of nature. Natural objects, dealing with stress applied by the environment have their specific ways of dealing with this stress (avoiding it or incorporating it in their growth) and the Laplacian used in Laplace, Helmholz, Poisson, Schrödinger or similar equations of mathematical physics, is key to the study of such problems, ranging from vibrating strings and drums to vibrating plates and elementary particles, in space or in space and time.

Dealing with stress, natural objects not only divert stresses, but can also incorporate stress into their structure. One example is the "flower" structure in the test of sand dollars; this shape is directly related as curvature to the overall shape of the sand dollar. In biology one of the most prominent examples are "the spots of the leopard", whereby genes and gene actions are spatially distributed. The break-through in biology and chemistry was Alan Turing's paper, *The Chemical Basis of Morphogenesis* [186]. In Turing's vision two coupled non-linear reaction-diffusion equations model the spatiotemporal movement of the concentration of two *morphogens*. The stable equilibrium of a system can be disrupted by a diffusion driven instability, when in a system there is influx and outflux of chemicals. This

Fig. 8.4 Patterns on Mollusk shells. Copyright Violet Gielis

Fig. 8.5 Asclepiads and the various patterns of color and epidermal hairs structures. Copyright Martin Heigan

instability turns into spatio-temperal patterning, resulting in the spots of dogs and leopards, the stripes of zebra's and tigers, the patterns on mollusc shells (Fig. 8.4) and the patterns and spots on flowers of *Asclepiads* (Fig. 8.5).

It is not surprising then that the Turing equations (reaction-diffusion) can be written as two Helmholtz equations [187]. The coupling is more intricate than in the case of plate vibrations, but the underlying structure is the same. This is also why we focused on the general solution of the Helmholtz equations, rather than purely plate vibrations. Solutions to Laplace equations are called harmonic solutions, those for Helmholtz equations are called metaharmonic solutions [188].

R-Functions Revisited

Earlier we used R-functions to combine domains, a beautiful combination of logic and geometry; one of Rvachev books is called *Methods of Logic Algebra in Mathematical Physics* (1974). The origin of R-functions however, is not in geometric modeling, but in approximation theory, in particular in solutions for boundary value problems [136-138]. In 1958 Kantorovic and Krylov had proposed that the product of two functions, on the one hand a real-valued function ω having specified numerical values (e.g. zero) as value on the boundary and positive within the boundary, and on the other hand an unknown function Φ that allows to satisfy the differential equation (exact or as approximation), could be used to represent the solution to that differential equation [189].

Solutions of a differential equation with these boundary conditions can be represented in the form $u = \omega\Phi$. This was shown as a general theorem, but how to construct such functions and how to generalize to other boundary value problems, was left as an open challenge. Rvachev proposed that R-functions could answer both questions. R-functions behave as continuous analogs of logical Boolean functions (but are broader), and functions could be constructed with the desired differential properties and potential fields. Rvachev's solution thus addressed both challenges suggested by Kantorovic' general theorem. R-functions are guaranteed to give solutions to the boundary value problem at hand.

One major advantage of defining geometric domains in this way is that the solutions can be computed in every point on and within the domain, in two and three dimensions. This avoids the need of meshing and spatial discretization of domains, surfaces and bodies; this has been implemented in a mesh-free modeling system in computer-aided design and engineering and has been called *semi-analytic geometry* [138]. The development for mesh-free methods is one of the holy grails of analysis; Fourier's methods worked for a circular disk and a few other domains, but the development of solutions for other domains relied on discretization and approximations. In the absence of exact or analytic methods (in the sense of Fourier), finite difference methods (e.g. finite elements) have been developed and arguably the development of all design and engineering software is a result of the lack of an exact method. R-functions provided such methods.

By using supershapes and our Fourier projection method we also obtain both geometric domains and solutions structures. We make use of Fourier's method on separation of variables, namely solutions are sought as $u(\rho, \vartheta) = P(\rho).\Theta(\vartheta)$. The $P(\rho)$ part is $[rR(\vartheta)]^m$ (with imposed boundary conditions, or ρ^m for the general solution) and the $\Theta(\vartheta)$ part is the Fourier series. Our Generalized Fourier series $\rho(\theta) = \sum_{k=0}^{q} \left(\rho_k a_k \cos\frac{m_k\theta}{4} + \rho_k b_k \sin\frac{m_k\theta}{4} \right)$ was defined earlier with the goal of developing k-type series as a compact descriptor (as a Pythagorean-compact fewnomial) in Fourier series style with k terms. When this moderation for each mth term of the expansion includes $a_k = A_m$ and $b_k = B_m$ and $\rho_k = [rR(\vartheta)]^m$ we have the solution of the Laplace equation. Consequently, we have two immediate applications of the usefulness of the Generalized Fourier Series: the description of domains and the solutions to boundary value problems with low order of expansion.

Rvachev functions were a solution to Kantorovich and Krylov' general theorem. We can now align Gielis curves and surfaces to such advances and theorems, generalizing Fourier's original method to any normal polar domain, or as solution to Kantorovich' Theorem, whereby the classical trigonometric functions and our shapes give both the functions and the solutions. Open challenges are to find solutions to various boundary value problems using classical Fourier methods on combined supershaped domains with different centers, including Riemann surfaces, and to generalize this to more complicated domains like knots and links, via Generalized Möbius-Listing surfaces and bodies.

Chapter 9
Natural Curvature Conditions

What astonishes and bewilders me most is the scale on which Nature uses the Superformula. The search for mathematical formulas that can describe biological forms is an old one. The Gielis Formula makes it possible to describe an almost unimaginable number of natural shapes in a very simple way. One of the next challenges will be to analyze which mechanism allows organisms to use this formula for shaping its organs and bodies. And especially what we can learn from it. I am convinced that the use of this formula will also lead to an avalanche of technological breakthroughs and practical applications.

Tom Gerats

Curvature, the Central Notion in Geometry

Curvature, in all of its aspects is at the core of geometry. Going straight is central to geometry, but it is also a fiction, since we live in a curved world with gravitational attraction in landscapes of mountains, valleys and planes. Nevertheless it is a very useful fiction, and the basis of Euclidean geometry and modern science. How then to define being curved, intrinsically and/or as a deviation from planarity? This is one of the central questions in mathematics and geometry; we have derived Euclidean geometry, one geometry that is in accordance with our intuition and our position on earth (with its gravitational field). René Thom: "*Classical Euclidean geometry can be considered as magic; at the price of minimal distortion of appearances (a point without size, a line without width) the purely formal language of geometry describes adequately the reality of space*" [9].

If we ride our bike on a long, straight and flat road, we can maintain a constant speed. When the road turns, the force we experience is directly related to the curvature of our turn; force F is acceleration a times a constant (in this case mass, $F = m \cdot a$) and acceleration is curvature of the trajectory. We may need to slow down.

© Atlantis Press and the author(s) 2017
J. Gielis, *The Geometrical Beauty of Plants*,
DOI 10.2991/978-94-6239-151-2_9

If no force had been applied, we would have continued in a straight line, eventually drifting off into space in an ideal world. A traveller along a curve or path risks at any point going straight, along the tangent line, which can be defined analytically in each point P. The traveller is kept on the path or traces out a path due to external forces; one example is a projectile following a parabolic curve. Curvature going beyond P is then defined by how much the path deviates from the tangent in a point P. The deviation from this line is measured as an angle between curve and tangent.

The curvature of curves in the Euclidean plane was determined analytically around 1670 by Isaac Newton, via osculating circles. Newton showed analytically that in each point of the curve, a circle can be applied, which is a direct measure of the deviation from this line. The radius R of the osculating circle gives this measure, and curvature itself is the inverse $\kappa = \frac{1}{R}$. If the curve makes a sharp turn, the curvature is high; if the turn is gentle, curvature is low. How the circle is applied to the curve is also well-defined: the normal direction, which contains P and the centre C of the oscullating circle, is perpendicular to the tangent line. If we know the tangent, we automatically know along which line the centre of the circle of curvature is to be found. When we know not only the curvature in one point, but of the whole curve, we have the fundamental theorem of plane curves in Euclidean geometry: *The knowledge of the curvature completely determines the curve.* Stated differently, curves with the same curvature are the same, and differ only in position [190].

Moving in Space

In geometry the definition of curvature is in the Oresme tradition a point-wise, local operation and this can also be used for curves in space. In each point of a curve in a plane a best fitting circle can be fitted, and this is called the osculating circle, and the orientation of the circle is defined by the normal, perpendicular to the tangent, and this circle is in the osculating plane [191]. For any point on a 3D curve, we can also define a tangent in space, and also here the direction of the osculating circle can be defined by the perpendicular direction to the tangent (Fig. 9.1).

In space this plane is one of the three planes, the other being the rectifying plane R spanned by $\{t, b\}$ and the binormal plane, spanned by $\{n, b\}$. The binormal vector is perpendicular to the tangent and the normal vector. In this sense the binormal defines a 'third dimension', so that in each point P we can fit an XYZ coordinate system, the orientation of which is fixed by the direction of the tangent and normal. In fact the Serret-Frenet-Pagani equations give the exact relations

$$t' = \kappa n, \quad n' = -\kappa t + \tau b, \quad b' = -\tau n \qquad (9.1)$$

These equations define *curvature* κ as the ratio $\frac{t'}{n}$ and *torsion* τ as the ratio $-\frac{b'}{n}$ with t' and b' denoting the change in t and b with respect to arc length. Both κ and τ

Fig. 9.1 Tangent, normal
and binormal vectors and
planes

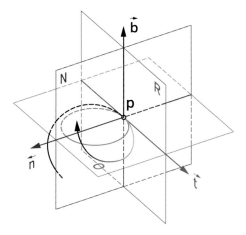

are measures of curvature. In the early history of the curvature of curves, κ and τ were called *"double courbure"*. The fundamental theorem for planar curves can be extended: *Knowledge of κ and τ completely determines a 3D curve.*

They are connected through the Darboux vector. If one moves through space on a curve with s the time parameter, then the moving frame $\{t, n, b\}$ moves in accordance with the Serret-Frenet-Pagani equations. This motion contains, apart from an instantaneous translation, an instantaneous rotation with angular velocity vector given by the Darboux vector $d = \kappa b + \tau t$. The direction of the Darboux vector is that of the instantaneous axis of rotation, and its length $\sqrt{\kappa^2 + \tau^2}$ is called the angular speed [191].

$$t' = d \times t, \qquad n' = d \times n, \qquad b' = d \times b \qquad (9.2)$$

Moving on Curves on Surfaces

On curves one has considerable freedom to move. For surfaces to be coherent, the double curvatures reduce to the maximum and minimum curvature κ_1 and κ_2, respectively. These are the maxima and minima of an infinite number of curvatures.

For surfaces the situation was clarified due to the work of Euler and Meusnier, who developed the notion of principal curvature. The principle is very similar to curves, but now we study the tangent plane in a point P on a surface; magic starts to happen. First, we can draw infinitely many tangents through the point P since in space there are an infinite number of directions. Since all these tangents lie in one plane, all normal directions, defined as lines perpendicular to the tangent, coincide. The curvature in all directions will be different in general (unless the surface is a sphere) but all radii of curvature lie along the normal direction, and all centers of all the osculating circles, lie on that normal. Among these radii of curvature, there will be a maximal value $\kappa_{max} = \kappa_1$, and a mimimal value $\kappa_{min} = \kappa_2$. Now comes the real magic: one of the most amazing results in geometry and mathematics is that for surfaces the maximal and minimal curvatures are in perpendicular normal planes! (Fig. 9.2).

At a point P we can draw an infinite number of planes having the normal as common line. In each plane a curve can be fitted to point P with curvature of that curve in the point P defined in the sense of Newton's osculating circles. There are two distinct planes, one with the maximal curvature κ_1, the other with the minimal curvature κ_2 and amazingly, they turn out to be perpendicular. This is one of the true gems of geometry (one of many). All other curvatures κ_n in P in any (normal) direction can be determined from κ_1 and κ_2, since these are perpendicular: $\kappa_n(\vartheta) = \kappa_1 \cos^2\vartheta + \kappa_2 \sin^2\vartheta$. So one can draw an ellipse of curvature to visualize the various values of κ_n.

Once we know the principal curvatures, they can be combined via the classical means of the Ancient Greek. The arithmetic mean of $\kappa_{max} = \kappa_1$ and $\kappa_{min} = \kappa_2$ gives the mean curvature $H = \frac{\kappa_1 + \kappa_2}{2}$ and the square of the geometric mean is the Gaussian curvature $K = \kappa_1 . \kappa_2$. The Gaussian curvature is technically defined as the ratio of a surface area of a small parallelogram drawn on that surface to the image of the area developed on a sphere. This is in line with Newton's curvature for a curve, based on the approximation with an osculating circle. It is also pointwise, but now approximated with a sphere; as the areas converge to a point, the limit of the ratios of these area is equal in value to the product of the principal radii.

In this way differential geometry (geometry in the small) became fully embedded in mainstream mathematics, giving rise to various definitions of curvature, relations between curvatures, Laplacians etc., all central in the application of mathematics in the natural sciences. One remark (and we refer to Laplacian, and the remark of Schrödinger): technically only second order contributions of curvature are taken into account and higher order contributions are neglected.

Another important remark is that the experience of curvature in the above sense does not always correspond with our intuition and perception. Going with or against the wind, or climbing a mountain or sliding downward, or anything anisotropic (at least from our point of view) our experiences will differ. These are challenges

Fig. 9.2 Curvature of surfaces with normal, maximal and minimal curvatures

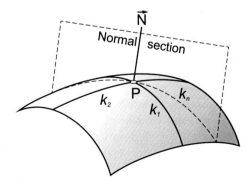

which require new points of view, like Minkowski and Finsler geometries, based on Lamé curves or on Lamé curves or Gielis transformations.

Notwithstanding this, everything works excellent: we have a straight line, connected with tangents, as first derivative, and we have the circle, connected with curvature and the second derivative. These are the main tools for our study of natural shapes and phenomena. Almost everything of importance in geometry can be reduced to the study of the relations between these tools. One example which we will discuss further is the angle between position vector and tangent vector, which in natural shapes tends to be constant.

K and H are the most commonly known measures, but to describe shape, two other measures of curvature (or curvedness) are also extremely useful. First is the Casorati curvature C, the arithmetic mean of the squares of principal curvatures [192]:

$$C = \frac{\kappa_1^2 + \kappa_2^2}{2} \tag{9.3}$$

The rationale for this curvature is some strange consequences of K and H. A cylinder, definitely curved in our eyes, has Gaussian curvature K zero since the minimal curvature is zero (and hence the product). This holds for all developable surfaces; fold a flat piece of paper into a cylinder or a cone, and the Gaussian curvature is still zero, because one of the principal curvatures is zero. The mean curvature H is not perfect either to describe curvedness: for minimal surfaces (e.g. a clearly curved catenoid) H is zero. So, while Gaussian or mean curvature vanishes for surfaces that look intuitively curved (like a cylinder), the Casorati curvature C only vanishes at planar points. This is not a perfect solution either since the Casorati curvature of sphere and catenoid is the same, as Eugène Catalan remarked in a letter (1891) to Felix Casorati [193]. Indeed, in catenoid and sphere $\kappa_1 = \kappa_2$ but in a catenoid κ_1 and κ_2 have opposite signs. The square root of the Casorati curvature has been proposed as a measure of *curvedness* [194].

The Shape Index σ of Jan Koenderink and Andrea van Doorn, is a really excellent way of describing how surfaces are curved in space, with a beautiful geometric meaning [194].

$$\sigma = \arctan\left(\frac{\kappa_1 + \kappa_2}{\kappa_1 - \kappa_2}\right), \tag{9.4}$$

Natural Curvature Conditions and Ideal Submanifolds

For mathematics and its applications we can then search for optimal relations between curvatures or extremals of these scalar-valued characteristics. From the point of view of optimal relations, we have already mentioned the inequality

$K \leq H^2$. Minimizing H or H^2 is one way to strive for equality in the classical inequality $K \leq H^2$ as one of the driving forces of nature, with equality $K = H^2 = 0$ attained in minimal surfaces, as in catenoids and planes. It is important to understand the product of principal curvatures K as an *intrinsic* characteristic, whereas H is an *extrinsic* characteristic of shape.

Inequalities give a dynamic character: in soap bubbles, the environment is always changing, and microscopically the soap molecules continuously adapt their configurations to comply with these changes and keep some constant relation between inside and outside pressure. This can be noticed by the continuously changing thickness of the bubble, accounting for the colorful patterns of soap bubbles, and how they move across the surface. The equality is not a given, it is a continuous dynamic process of striving towards or optimizing configurations. All in all however, the inequality $K \leq H^2$ is nothing but the inequality between the square of the geometric mean, and the arithmetic mean of the principal curvatures, a beautiful application of the relations that the Greek discovered between numbers in a arithmetical and geometrical way. Unlike for positive numbers only where the inequality is strict $GM < AM$, curvatures can be negative and equality can be achieved. The geometrically natural curvature conditions study, in their simplest form, the *constancy of curvatures* like the extrinsic curvature, expressed by the *mean curvature H* (expressing uniform surface tension) and the intrinsic curvatures, expressed by *Gauss curvature K* for surfaces M in Euclidean 3D-space.

The other strategy is finding extremals in curvatures (intrinsic or extrinsic). Finding extremals of H^2 is a natural choice, embodied in the Willmore functional: $\int H^2 dA$ on closed surfaces in 3 dimensional space. These extremals can then be used in inequalities: the Willmore Theorem states that this functional $\geq 4\pi$ with equality (and thus a minimum) for the round sphere. The natural problem is to find out what the minima are when the surfaces have additional constraints, topologically or metrically [195]. This finds applications in biology as the Helfrich energy for membranes. The cell membrane is the boundary between a living cell and its environment. It is a surface connecting the inner with the outer and so we search for optimal relations between *outer* and *inner* characteristics of shapes. Instead of using the mean curvature, it uses H^2, which resonates with what Casorati did in 1890: square the curvatures. This becomes clear when we expand the Willmore conditions:

$$H^2 = \left(\frac{\kappa_1 + \kappa_2}{2}\right)^2 = \left[\left(\frac{\kappa_1^2 + \kappa_2^2}{4}\right) + \frac{\kappa_1 . \kappa_2}{2}\right] \qquad (9.5)$$

or:

$$H^2 = \frac{\left(\frac{\kappa_1^2 + \kappa_2^2}{2}\right) + (\kappa_1 \kappa_2)}{2} \qquad (9.6)$$

H^2 then simply is the arithmetic mean of the Casorati curvature $\frac{\kappa_1^2 + \kappa_2^2}{2}$ (a measure of extrinsic curvature) and the Gaussian curvature $\kappa_1 \cdot \kappa_2$ (a measure of intrinsic curvature) of a surface. Using the squares of the curvatures, we find that their geometric mean is $\sqrt{(\kappa_1 \kappa_2)^2} = |K|$, and:

$$H^2 = \frac{\left(\frac{\kappa_1^2 + \kappa_2^2}{2}\right) + \sqrt{(\kappa_1 \kappa_2)^2}}{2} \tag{9.7}$$

Having arithmetic and geometric means, another natural question is whether we can apply the graphical methods described earlier to determine *AM* and *GM* (indirectly via *HM*), using k_1 and k_2. The Harmonic Mean has been used recently in the concept of amalgamatic curvatures. Since we are dealing now with the squares (directly in *AM* with the Casorati curvature, with positive curvatures only in *GM*), we are in need of a different spacing, a quadratic one. In order to align this we need to map a parabola onto a line. This can be achieved by using parabolic trigonometric functions.

So in many ways we find inevitable relations between curvatures that fundamentally relate to the *intrinsic nature* of these shapes, and curvatures that fundamentally relate to the shape, which these shapes assume in their ambient world, i.e. *extrinsic curvatures*. A main goal of differential geometry is to describe and understand the shape of surfaces (and generalizations of shape to manifolds). We then study these shapes in Euclidean space (or generalization), as submanifold in a Euclidean space. Minimal surfaces for example are submanifolds realizing the equality $K = H^2 = 0$.

In general terms, a submanifold is said to be *an ideal submanifold* in some space when it actually realizes at all of its points the equality in such generally holding inequality. Inequalities give a dynamic character. Submanifolds have intrinsic structures, which normally do not change. Ideal submanifolds thus experience the least possible external curvature that might be imposed on them by the form they assume in a given surrounding world. This is akin to experiencing the least possible stress that the living conditions in these ambient spaces may impose on the submanifolds, the creatures, which happen to live in these worlds. An ideal submanifold is then *ideally embedded* [196]. The most well known example is the spherical shapes of soap bubbles, but also space-time models of the FLRW-type have been shown to be ideally embedded.

Similarly, plant shapes in morphology and anatomy, crystals and snowflakes, also strive for the status of *ideal submanifold*, adapted to the surrounding space in a best possible way, in a certain type of equilibrium, irrespective of development times or materials. One will not fail to see the inevitable connection to the natural world in evolution, and how the intrinsic qualities of shapes, represented by either or both biological or mathematical DNA, necessarily have to relate to and interact with the world-around. All creatures, great and small, animate or inanimate, have to cope with a range of challenges in different ways. Coping with the "environment" successfully means nothing more or less than trying to find a best way of living in a best world, since both the fit creature, and the world around, play their part in this interactive and inevitable co-existence and co-evolution. This may involve various contributions from the extrinsic and intrinsic curvatures, as arithmetic mean, or weighted arithmetic mean.

Constant Mean Curvature Surfaces and Conic Sections

How to measure the stress on a surface in its environment? Via mean curvature H. The Laplacian is directly related to the mean curvature H on surfaces. For surfaces in n dimensions it is, according to a theorem of Beltrami:

$$\Delta x = -nH \tag{9.8}$$

In this way the stress on a surface can be conceptualized using the principal curvatures of the surfaces k_1 and k_2, in particular their arithmetic mean H. This is a nearly universal principle in physics, and biology as well. Minimizing H or H^2 is one way to strive for equality in the classical inequality $K \leq H^2$, between the intrinsic or Gaussian curvature K and the square of the mean curvature H. It is a geometric version of the classical inequality $\sqrt{GM} \leq AM$. There are a number of options. The stress on the surface is distributed as evenly as possible and can be zero or larger than zero. There are two general cases in which $H = 0$: the plane and the catenoid. The latter is the shape obtained when two rings are pulled out of a soap solution. The profile curve between the rings is a catenary, and revolving this around a central axis, yields the catenoid. In both plane and catenoid, the inequality becomes an equality in the sense that $K = H = 0$. In soap bubbles spheres are the perfect shape for a free-floating solution, minimizing (differences in) surface tension, but the mean curvature is not constant.

There are three other cases, in which $H \neq 0$, but distributed as evenly as possible. This is the case in the cylinder, the unduloid and the nodoid. All these surfaces are generated from profile curves revolved around a central axis. All such shapes are known as constant mean curvature CMC surfaces, or Delaunay surfaces. The *rotational surfaces of Delaunay*, i.e. those of constant mean curvature H, (planes and catenoids for $H = 0$, spheres and circular cylinders and unduloids and nodoids for $H \neq 0$) are of very special interest, since their profile curves or roulettes, are obtained by rolling conic sections over a plane (Table 9.1). When the curves obtained by rolling a circle, ellipse, parabola and hyperbola are rotated around the base line, they indeed give rise to cylinders, unduloids, catenoids and nodoids respectively, another of Geometry's Fantastic Gems. Delaunay shapes effectively occur in various living organisms, feeling at home in the sea. Once more conic sections, based on the application of areas of the Pythagoreans, and physically exemplified by light bundles through a pinhole, are involved. The studies of

Table 9.1 From conic sections to CMC surfaces

Conic section	Focus	Profile curve	CMC surface
Circle	Center	Line	Cylinder
Ellipse	Focus	Sinuous line	Unduloid
Parabola	Focus	Catenary	Catenoid
Hyperbola	Focus	Elastic curves	Nodoid

minimal surfaces, known as Plateau's Problem, have occupied the minds of the brightest mathematicians and physicists in the past two centuries.

Miyuki Koiso and Bennett Palmer have defined Constant Mean Anisotropic Curvature CAMC surfaces, [13, 197, 198] i.e. anisotropic analogues of catenoids and Delaunay surfaces (Fig. 9.3). According to Bennett Palmer:

> The nearly universal principle in the natural sciences is that the equilibrium configuration of a system can be found by minimizing its total energy among all admissible configurations. When we are considering the surface interface between two or more immiscible materials, the surface geometry is determined by minimizing the surface tension subject to whatever additional constraints are imposed by the environment. These constraints may take the form of boundary conditions or may include constraints that the volume or even the surface area be preserved under deformations. In addition, there may be additional energy contributions depending on the boundary such as wetting and line tension. There may also be energy contributions arising from external forces such as gravity. For materials, which are in an ordered phase, the interfacial energy may be anisotropic, i.e. its density may depend on the direction of the surface. This is particularly true of the surface if a crystal, a fact that is empirically obvious from the way that the surface geometry tends to favor certain directions. There is a canonical equilibrium surface, called the Wulff shape that can be characterized as the absolute minimizer of the free energy F among all surfaces enclosing the same three dimensional volume as the Wulff shape.

A Wulff shape is the "sphere" for an anisotropic energy in the sense that it is the minimizer of the energy for a fixed volume. More generally, from a technical point of view, any of the three dimensional shapes defined by Eq. 5.17 or similar is a Wulff shape which means that the shapes are minimizers for a certain *anisotropic* energy functional. This gives rise to supercatenoids, superunduloids and supernodoids. The examples that they used for Wulff shapes are Lamé surfaces or supershapes. It will come as no surprise that such shapes also occur in nature.

The *supercatenoid* has the property that sufficiently small pieces of it minimize the anisotropic energy defined by the Wulff shape among all surfaces having the same boundary. Sufficiently small pieces of the unduloid and nodoid minimize the anisotropic energy among all surfaces having the same boundary *and* enclosing the same three-dimensional volume (Fig. 9.3).

This anisotropic view provides a geometrical foundation for the development of natural shapes, with a direct relation to soap films, minimal surfaces and surfaces of constant mean curvature, but extended to CAMC for anisotropic energies. Research of supershapes with the remarkable correspondence to natural shapes will increasingly become the study of the way they are solutions to optimization problems through their properties, curvatures (the mathematical DNA of shapes) and related invariants.

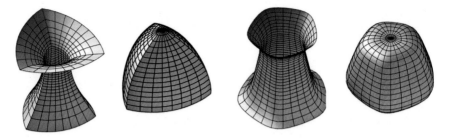

Fig. 9.3 Examples of Wulff shapes and associated catenoids. Copyright Bennett Palmer

Curvature Defined by Lamé and Gielis Curves

This extension to the anisotropic case is an important step in addressing René Thom's challenge, namely *"to construct an abstract, purely geometrical theory of morphogenesis, independent of the substrate of forms and the nature of the forces that create them"*. The descriptive part are curves and surfaces and the shapes which are derived via Gielis-transformations (of the conic sections), which *describe and determine* in a uniform and universal way an enormous diversity of natural shapes. Conic sections are at the core, once more.

Returning to conic sections, we know that Lamé curves have conic sections as special case. The question then is whether we can generalize the notion of curvature in a geometrical way, based on our uniform description. The definition of curvature is essentially a local operation. In each point of a curve a circle can be fitted and its radius is the inverse of the curvature. If however, one desires to describe the curvature of a square, then in each point of the flat sides with zero curvature, (intuitively) a circle with infinitely large radius needs to be applied; except in the vertices, where the curvature is infinitely large, and an infinitely small circle needs to be applied. The straightforward question, with the knowledge we have on Lamé curves, is then, whether it would not make more sense to define the curvature of a square on the basis of the osculating square itself, i.e. the corresponding Lamé curve. The square is a unit circle anyway, and one could choose the inscribed square for convenience and ease, avoiding limits and infinity. In a similar way the curvatures of starfish, pentagons and pentagrams are defined by the corresponding osculating supershapes.

We have to think intrinsically. After all, we think that vectors stretch whereas they might not stretch at all for internal observers. If we rotate the osculating starfish a few degrees, it will no longer be osculating globally (Fig. 9.4 left). This however is a strictly Euclidean view. The correct way to rotate the osculating circle is to follow the distance measures and symmetry dictated by the space, so that it will shorten and lengthen according to the angle (but we have this shortening and lengthening under full control). The osculating unit circle is precisely the shape

Fig. 9.4 Oscullating starfish and chains along the curve

(Fig. 9.4 center). This is easily understood if we consider a chain of beads on the curve: when the string of beads moves on the starfish curve, they will nicely follow it (just like a string of beads on a circle will follow the circle). The shortest path along the curve is the curve itself (Fig. 9.4 right). The triviality is that the shortest perimeter to enclose the curve is simply to follow the curve itself.

Like in the case of the circle that has precisely one osculating circle, curvature then becomes a global operation, at once describing the curvature of the complete shape, in full precision. In 3D the same procedure could be applied.

For a circle the curvature $\kappa_C = \frac{1}{R}$ whereby $\rho(\vartheta) = R$.

$$\sqrt{\cos^2\vartheta + \sin^2\vartheta} = \frac{1}{\rho(\vartheta)} = 1 = \kappa_C \tag{9.9}$$

Definition 9.1 Using Lamé curves as measure of curvature κ_L we obtain

$$\sqrt[n]{\cos^n\vartheta + \sin^n\vartheta} = \frac{1}{\rho(\vartheta)} = \kappa_L \tag{9.10}$$

Definition 9.2 Using Gielis curves as measure of curvature κ_G we obtain

$$\sqrt[n_1]{\left|\frac{1}{A}\cos\left(\frac{m}{4}\vartheta\right)\right|^{n_2} \pm \left|\frac{1}{B}\sin\left(\frac{m}{4}\vartheta\right)\right|^{n_3}} = \frac{1}{\rho(\vartheta)} = \kappa_G \tag{9.11}$$

Corollary 9.3 Piecewise linear curvature can be described piecewise by Gielis polygons or polygrams. For m a positive rational number:

$$\frac{1}{\varrho(\vartheta)} = \lim_{n_1 \to \infty}\left[\left|\cos\left(\frac{m}{4}\vartheta\right)\right|^{2(1-n_1\log_2\cos\frac{\pi}{m})} + \left|\sin\left(\frac{m}{4}\vartheta\right)\right|^{2(1-n_1\log_2\cos\frac{\pi}{m})}\right]^{1/n_1} \tag{9.12}$$

This allows us to study curvature of lines, polygons and irregular curves, in a piecewise way. Rather than a purely global or local operation, one could also consider a *glocal* operation, whereby to certain sections of shapes (arcs of a curve) a Lamé curve or supershape could be applied.

The Inverse Square Law Generalized

The curvature of a curve is the inverse of the radius of the osculating circle to the curve in a point $\kappa = \frac{1}{R}$, and this radius lies on the same line as the normal vector. From this many inverse square laws in physics are derived, defined by $\frac{1}{R^2}$. In general, the intensity of a signal from a point source decreases with distance as the square of the distance to the source. This is the case in light, radiation and sound, electrical charges (Coulomb's law) and gravitational attraction between two masses. One can rewrite these laws for masses and charge, respectively as:

$$\frac{F}{G} = \frac{m_1 m_2}{r_1 r_2} \quad \text{and} \quad 4\pi\varepsilon_0 . F = \frac{q_1 q_2}{r_1 r_2}$$

These laws immediately follows from the definition of a circle:

$\varrho(\vartheta) = \frac{1}{\sqrt{\cos^2\vartheta + \sin^2\vartheta}}$ can be rewritten as $\cos^2\vartheta + \sin^2\vartheta = \frac{1}{\varrho^2(\vartheta)} = 1$ which is the Pythagorean Theorem for the unit circle. For a constant function R, we get $\frac{\cos^2\vartheta + \sin^2\vartheta}{R^2} = \frac{1}{\varrho^2(\vartheta)}$ or $\frac{1}{R^2} = \frac{1}{\varrho^2(\vartheta)}$. In the very same way an inverse nth-power law based on Lamé curves can be defined as:

$$|\cos\vartheta|^n + |\sin\vartheta|^n = \frac{1}{\varrho^n(\vartheta)} \tag{9.13}$$

and for Gielis curves:

$$\left|\frac{1}{A}\cos\left(\frac{m}{4}\vartheta\right)\right|^{n_2} \pm \left|\frac{1}{B}\sin\left(\frac{m}{4}\vartheta\right)\right|^{n_3} = \frac{1}{\varrho^{n_1}(\vartheta)} \tag{9.14}$$

or in relation to functions and transformations of these functions:

$$\left|\frac{1}{A}\cos\left(\frac{m}{4}\vartheta\right)\right|^{n_2} \pm \left|\frac{1}{B}\sin\left(\frac{m}{4}\vartheta\right)\right|^{n_3} = \frac{f^{n_1}(\vartheta)}{\varrho^{n_1}(\vartheta)} \tag{9.15}$$

Do such inverse laws exist in nature? Yes, they exist and they are very general and come with the shape.

Scholium 9.4 $\frac{1}{\varrho^n(\vartheta)}$ and $\frac{1}{\varrho^{n_1}(\vartheta)}$ hold for the shapes they represent. The inverse square law is derived immediately from the definition of the circle and the Pythagorean Theorem. In the same way the inverse nth-power law follows immediately from the definition of Lamé curves and supershapes and the corresponding generalized Pythagorean Theorem $\left(\cos_p \vartheta\right)^p + \left(\sin_p \vartheta\right)^p = 1$. In other words, these laws hold for these curves by definition.

So we find that the definition of curvature (and consequences like the inverse square law) comes for free with the description of the shape in a generalized Pythagorean way.

Sectional Curvature of Riemannian Manifolds

Thus far we dwelled in two and three dimensions (or four, taking into account the two perpendicular planes with different axes). Generalizations to higher dimensions are possible, but we will limit to one result related to curvature.

The *Theorema egregium* of Gauss (1826) asserted that Gaussian curvature of surfaces is invariant under isometric deformations of surfaces M^2 in $\mathbb{E}^3$. It is the intrinsic characteristic *par excellence* and this theorem yielded the distinction between *intrinsic* and *extrinsic* curvatures. Riemann generalized this for higher dimensions and this led to the creation of Riemannian geometry, which became the core of modern differential geometry. Riemannian spaces are characterized by their curvatures and all curvature information is contained in the Riemann curvature tensor R. This means that the knowledge of curvature is the same as that of all sectional curvatures. Riemann showed that the sectional curvature is directly related to the Gaussian curvature, so that such sectional curvatures can be defined in the very classical Oresme-Newton tradition when embedded in a Euclidean space.

Earlier we referred to the *intrinsic* and *extrinsic* curvatures on surfaces, the first, Gaussian curvature, defined by the product of the two principal curvatures (the square of the geometric mean) and the second the mean curvature H, defined as the arithmetic mean of the two principal curvatures. The inequality $K \leq H^2$ then plays a key role in the shape that natural organisms, living or not, attain in a given environment, striving for equality in the inequality, as is the case for planes and catenoids [12]. Such relationships can be generalized to higher dimensions, in which a surface becomes an n-dimensional manifold, which can be embedded in a $n+m$ dimensional Euclidean space. The notion of n-dimensional manifold introduced by Bernhard Riemann, generalized the insight of Gauss on the curvature of surfaces. And it turned out that the same principles that apply for curves and surfaces, can be generalized for any dimension.

Theorem 9.5 *[199]: The sectional curvature $K(P)$ of any submanifold M^n of a Euclidean space $\mathbb{E}^{m+n}$, for any tangent plane section P at any of its points p, equals the Gauss curvature at p of the corresponding 2-dimensional normal section $\sum_p^2$ of M^n in $\mathbb{E}^{m+n}$.*

Theorem 9.6 *[199]: Any sectional curvature of a submanifold M^n in $\mathbb{E}^{m+n}$ is determined by the curvatures of at most two associated Euclidean planar curves.*

In this way the Oresme-Newton tradition comes full circle for any m-dimensional manifold in a $n+m$ dimensional space. Moreover, Theorem 9.6 can be reduced to the curvature of one associated Euclidean planar curve. From this we have the following:

Scholium 9.7 The curvature of this Euclidean curve can be determined in the classical way, using osculating circles, or by a Gielis transformation on the Euclidean circle.

Scholium 9.8 It is then possible to develop a geometry based on the notion of curvature, with osculating supershapes.

Curvature Diagrams

There is a graphical way of visualizing curvatures and their relations, via a plane graph with $X = k_1 = \kappa_1$ and $Y = k_2 = \kappa_2$, with κ_1, κ_2 the principal curvatures. Since $k_1 \geq k_2$ we need to consider only a halfplane. Within this halfplane the conditions $k_1 = k_2$, $k_1 + k_2 = R_H$, $k_1 . k_2 = R_K$, $k_1^2 + k_2^2 = R_C$, with R_H, R_K, R_C constants, are readily visualized as straight lines (constant mean curvature H), hyperbola (constant Gaussian curvature K) and circle (constant Casorati curvature), respectively. It is then easy to see, for given values of k_1, k_2 what various conditions on curvatures mean [200].

The curvature ellipse related to Euler's theorem describing the normal curvatures κ_N also fits here. Thinking back of the parabola and the application of squares and rectangles, a constant product $K = k_1 . k_2$ (the intrinsic quality of a surface) remains constant if the original rectangle with sides k_1, k_2 and area $A = k_1.k_2$, is turned into some other rectangle (along the hyperbola, for example with sides $3k_1$ and $\frac{1}{3}k_2$) or square (with sides $\sqrt{k_1.k_2}$) with the same area. In other words, for the intrinsic quality $K = k_1.k_2$ to remain constant, we can find different realisations with different k_1, k_2, as long as they scale in such a way to keep their product constant. We can do this not only for 2D surfaces, but also for manifolds M^n in $\mathbb{E}^{m+n}$ since the sectional curvatures are directly given by the Gaussian curvatures of the 2D normal sections.

One could easily generalize the relation between the principal curvatures to Lamé curves $k_1^n + k_2^n = R_L^n$. The Casorati curvature, as a circle, is then $C = \frac{1}{2}R_L^{n=2}$

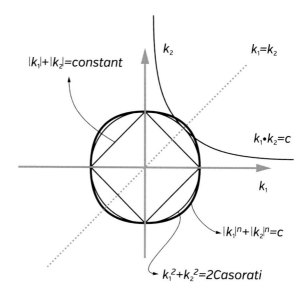

Fig. 9.5 Curvature diagrams for constant mean, Gaussian, Casorati and Lamé curvature

and the mean curvature $H = \frac{1}{2}R_L^{n=1}$. Note that with constant Lamé curvature we operate directly via the principal curvatures, which is different from κ_L defined earlier on the basis of osculating Lamé curves. We are free to apply Gielis transformations on these graphs (Fig. 9.5).

Flowers as Analytical Method for Studying Curvature

A further advantage compared to local operations with circles, is that this *glocal* or *global* curvature can be expanded in a finite polynomial. One of the main tools in differential geometry and analysis is the expansion in Taylor series, due to Newton and by definition infinite, but we have a tool for finite expansions available in the form of Chebyshev polynomials. This may be understood when we consider superparabola's of the type $y = |x^n|$. These can also be used as measures of curvature, in a local (pointwise) or glocal (section of a curve) operation, or even the global operation if the curve under study is $y = |x^n|$. The oligomial in terms of Chebyshev polynomials defining $x^n = \frac{1}{2^n}\binom{n}{n/2} + \frac{1}{2^{n-1}}\sum_{k=0}^{\frac{n-1}{2}}\binom{n}{k}T_{n-2k}(x)$. This is a precise expression and avoids the need for infinite series. It coincides also with the property of Chebyshev polynomials of best-fitting approximation of a curve.

In the tradition of Oresme of defining curvature by using a circle, the Pythagorean Theorem provided the compact expression of the curvature. Newton built upon this and determined the curvature of a curve analytically by infinite series, still essentially a local operation. Using transcendental functions—trigonometric ones—Fourier succeeded in a global description of a curve, in terms of infinite series, Fourier series. Now we use the compact (Pythagorean style) expression of a supershape or Lamé curve to define curvature, as well as a finite expression using Chebyshev polynomials. Obviously this coincides with Fourier cosine series, since the correspondence with Chebyshev polynomials is complete. In this way we combine the classical views of Oresme-Newton-Fourier (and in fact Ptolemy) into a more general notion of curvature beyond a local operation only.

In the beautiful landscapes and universes of geometry, one might easily get lost, but shortcuts can always be found. Euler's Theorem on the curvature in a point on a surface, written in terms of principal curvatures, can be rewritten as follows [201]:

$$\frac{1}{\rho_n} = \frac{1}{R_1}\cos^2\vartheta + \frac{1}{R_2}\sin^2\vartheta = \frac{1}{2}\left(\frac{1}{R_1} + \frac{1}{R_2}\right) + \frac{1}{2}(\frac{1}{R_1} - \frac{1}{R_2})\cos 2\vartheta \qquad (9.16)$$

or,

$$\kappa_n = \kappa_1\cos^2\vartheta + \kappa_2\sin^2\vartheta = \frac{1}{2}(\kappa_1 + \kappa_2) + \frac{1}{2}(\kappa_1 - \kappa_2)\cos 2\vartheta \qquad (9.17)$$

$$\frac{\partial\kappa_n}{\partial\vartheta} = (\kappa_1 - \kappa_2)\sin 2\vartheta \quad\text{and}\quad \frac{\partial^2\kappa_n}{\partial\vartheta^2} = 2(\kappa_2 - \kappa_1)\cos 2\vartheta \qquad (9.18)$$

κ_n and its derivatives can be written in terms of the principal curvatures and $\cos 2\vartheta$ and $\sin 2\vartheta$ (double angles, so related to chords), easily visualized in curvature diagrams. The ratio $(\kappa_1 + \kappa_2)$ and $(\kappa_1 - \kappa_2)$ appears in Shape Index $\sigma = \arctan\left(\frac{\kappa_1 + \kappa_2}{\kappa_1 - \kappa_2}\right)$. Obviously, $\cos 2\vartheta$ and $\sin 2\vartheta$ can be rewritten as T_2 and $\sqrt{1 - x^2}U_1$ respectively, so that also classical curvatures can be expressed directly as Chebyshev polynomials. Our generalized curvatures can then be expressed in this very same way. It started with a letter to Leibniz, and transformed into compact and powerful tools.

Circles and Spirals

Dealing with anisotropic shapes, we must take into account that what looks anisotropic to us, may well be isotropic for the shape itself. The example of the string of beads rotating round a supershape clarified this. In this case, intrinsic curvatures based on supershapes, are proposed. However, because of the compact structure and the trigonometric functions, we can relate it immediately to classical notions based on the circle. In traditional geometry, the geometry developed using our intuition of space, circle and straight line are the main tools, also in curved spaces, where differential geometry strives towards studying curvatures with Euclidean geometry, as much as possible. Now, we will show in a purely geometrical way how, in the study of natural sciences, most elementary shapes can be reduced to two basic shapes, the circle and the spiral. Applying Gielis Transformations to these two basic shapes then become strategic tools in the study of natural shapes and phenomena.

The ubiquity of circle and spiral as basic shapes or movements are not new. Goethe lived and worked in the same period as Gauss, also a student of nature, but in a different way. At the end of the 20th century, Goethe's "*Alles ist Blatt*" ideas were corroborated in the field of plant molecular biology. All manifold realizations of the concept *Leaf* are in principle variations on a theme, and a chain of molecular events, dependent on timing and position, is set in motion to produce vegetative leaves, bracts, spines, tendrils, or any of the flower organs, sepals, petals, pistils and stamen. For Goethe growth was expansion and contraction: growing stems and branches make the plant expand, and in the generative phase, everything is contracted, into very compact structures, governed by intrinsic parameters (internal clocks, cells, organs, DNA, biochemistry and physiology), in a synchronous dance with the environment (seasons, gravitation, light and nutrients…).

Towards the end of his life, around 1729–1831, Goethe, was particularly inspired by the spiral and its development. In *Über die Spiraltendenz der Vegetation*, he describes observations of many spiral tendencies in nature, in growing stems, flowers, inflorescences [202]. Of course, Goethe was not the first to refer to this and he cites and compares many published works on this matter. This spiral tendency would later become one of Charles Darwin's main interests. It is not widely known that Darwin gathered a lot of empirical evidence on plants (more than on Galapagos finches in any case). Among his main interests was the behavior of climbers (with spiral or helical tendencies) and rapid movements in plants.

So now there are two main tendencies: the upright growth and branching on the one hand, and the spiral growth of phyllotaxy and the structure within the flower (either spiral or whorled configurations) on the other. We can characterize both tendencies geometrically; the vertical tendency with the *circle* as basic figure, and aligning with the earth's gravitational field, versus the constant ratio with the development of leaves and organs from a *spiral* sequence on a conical meristematic dome (Fig. 9.6).

Fig. 9.6 Circular and spiral tendencies in developing flowers and inflorescences. Copyright National Botanic Garden Belgium

Going with and Going Against the Flow

In the classical approach with the Serret-Frenet-Pagani frame we study the double curvature (curvature and torsion) along the path with the tangent. But since we start from a center, we also have the availability of a position vector from a central point. In this case, we can study the relation between the tangent space and the position vector. For a curve given by $X(\vartheta) = (\rho(\cos \vartheta), \rho(\sin \vartheta))$, the second derivative $X''(\vartheta)$ describes the tension in the curve [165].

Now there are two possible opposites. Either position vector and tension vector are parallel $(X \parallel X'')$, or they are orthogonal $(X \perp X'')$. The parallel case $X \parallel X''$ occurs when the *vector* product of both vectors is zero. Then the basic figure is the *circle*. If the *scalar* product is zero we have the orthogonal case: $X \perp X'' \Leftrightarrow X.X'' = 0$. The basic figure is then the *spiral* (logarithmic spiral). So we have the two basic figures in nature in one coherent picture. The beauty is that both circle and spirals occur in flowers and flower development, so this is a perfect area of study (Fig. 9.6).

In this relation between X and X'', the tension in the curve can be aligned in two opposite ways. In the parallel case with circle as a basic figure the dynamic can be understood as *going with the flow,* and in the case of the spiral, the dynamic is to completely *oppose the flow.*

D'Arcy Thompson gave a beautiful description of how growth opposing the flow works:

> *In mechanical structures, curvature is essentially a mechanical phenomenon. It is found in flexible structures as a result of bending, or it may be introduced into construction for the purpose of resisting such a bending-moment. But neither shell nor tooth nor claw are flexible structures; they have not been bent into their peculiar curvature, they have grown into it.*

> *We may for a moment, however, regard the equiangular or logarithmic spiral of our shell from the dynamic point of view, by looking at growth itself as the force concerned. In the growing structure, let growth at a point P be resolved into a force F acting along the line*

Fig. 9.7 Spiral and D'Arcy Thompson

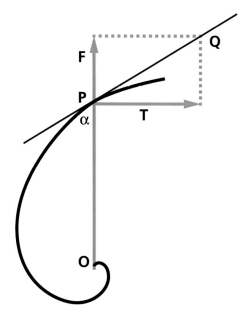

joining P to a pole O, and a force T acting in a direction perpendicular to OP; and let G be the magnitude of these forces (or of these rates of growth) remain constant. It follows that the resultant of the forces F and T (as PQ) makes a constant angle with the radius vector. But a constant angle between tangent and radius vector is a fundamental property of the "equiangular" spiral: the very property with which Descartes started his investigation, and that which gives its alternative name to the curve (Fig. 9.7).

In such a spiral, radial growth and growth in the direction of the curve bear a constant ratio to one another. For, if we consider a consecutive radius vector OP', whose increment as compared with OP is dr, while ds is the small arc PP', then dr/ds = cos α = constant. In the growth of a shell, we can conceive no simpler law than this, that it shall widen and lengthen in the same unvarying proportions: and this simplest of laws is that which Nature tends to follow. The shell, like the creature within it, grows in size but does not change its shape; and the existence of this constant relativity of growth, or constant similarity of form, is of the essence, and may be made the basis of a definition, of the equiangular spiral.

The position vector (also known as radius vector or location vector) is central in geometry [203]. It is the most elementary and natural geometric object giving the position relative to a reference point. This coincides completely with developments in contemporary differential geometry. The constant ratio spiral is the 2D version of *constant ratio submanifolds* of Bang-yen Chen in *n*-dimensions, with a constant ratio between the tangent and the normal direction on a submanifold. *D'Arcy Thompson's Law of Growth*, as it has been baptized by geometers [165], is a special case. This constant ratio is very important in natural shapes. Both circle and spiral are examples of constant ratio curves. For a circle this is obvious: the tangent and the normal are perpendicular going around the curve. For the spiral, growth is from a central point, and occurs at the end of the curve. The logarithmic spiral as we know, is characterized precisely by having a constant ratio between these two as well, namely between the resolvents of growth in length and width.

It also underlines the value of D'Arcy Thompsons magnificent book *On Growth and Form*. In this book he summarized much of the knowledge in geometry and how it could be applied to natural shapes. In 1917, half a century had elapsed since Darwin's Theory of Evolution was published, and around the turn of the century, Hugo DeVries had rediscovered Mendel's experiments. Feeling that this would lead to a strong divergence between physics and biology, D'Arcy Thompson wanted to emphasize the role of physical laws, mechanics and mathematics. In *On Growth and Form*, logarithmic spirals, allometric laws, constant mean curvature surfaces in marine organisms and much more, apply findings of Laplace, Delaunay and many other great mathematicians, to biology.

D'Arcy Thompson wanted to go against the trend of biology overemphasizing evolution as fundamental determinant in shapes. In this sense he applied his own law going against the flow. For him everything in nature, which can be studied, is subject to mathematical and physical laws. Thompson wrote: *"An organism is so complex a thing, and growth so complex a phenomenon, that for growth to be so uniform and constant in all the parts as to keep the whole shape unchanged would indeed be an unlikely and an unusual circumstance. Rates vary, proportions change, and the whole configuration alters accordingly."* It is an illusion to think that a molecule like DNA is the only or major determinant. As contemporary molecular biology has understood, development is a very closely regulated inter-play between internal (DNA, biochemisty, cells and organs) and external charac-teristics (soil, weather, season), via epigenetic imprinting. This aligns with contemporary developments in geometry, with circle and spiral as basic shapes. DNA itself for example is a Generalized Möbius Listing body with torsion [116] (in planar projected view a rational Gielis curve) with internal structure. The tension imposed globally or locally onto the molecule can be modified by changing its conformation locally or globally, transcription as one example. Transposable ele-ments may play an important role comparable to tuning a violin or a piano.

A basic shape as the circle is the basis for many of the optimal curves that have been studied throughout the ages. The spiral is associated directly to natural growth and the exponential function and many other tendencies. For example, Newton found that if the motion in a central force field obeys an inverse-cube law, the mass particle would spiral off into space, instead of revolving around in orbits defined by conic sections. The trajectory of a mass particle subject to a central force of attraction located at the origin, which obeys the inverse-cube law, is a curve of constant-ratio.

Applying Gielis Transformations to the Basic Shapes

Beyond circles and spirals, how do we deal with all types of anisotropies in nature? Most mollusks are not *Nautilus* shells or snails, the prime example of spirals in nature. With this analysis of constant ratio submanifolds, we can return to the opening statement of this book: "*The geometrical description of curves and surfaces, and the shapes that are derived* via *Gielis-transformations, describe and determine in a uniform and universal way an enormous diversity of natural shapes*" [3]. Indeed, we can simply apply Gielis transformations to the basic forms that naturally form in Euclidean geometry [165] (Fig. 9.8).

With Gielis transformations, we have a position vector with variable length, the variation given by the superformula, and this can be applied to circle and spiral. We already showed many examples of this. It is basically modifying the "curvature" with Gielis transformations, so we go from circle to the generalized form:

$$r = R \rightarrow \rho(\vartheta) = \frac{1}{\sqrt[n_1]{\left|\cos\left(\frac{m}{4}\vartheta\right)\right|^{n_2} + \left|\sin\left(\frac{m}{4}\vartheta\right)\right|^{n_3}}}.R \qquad (9.20)$$

and from spiral to its generalized form:

$$r = e^{k\vartheta} \rightarrow \rho(\vartheta) = \frac{1}{\sqrt[n_1]{\left|\cos\left(\frac{m}{4}\vartheta\right)\right|^{n_2} + \left|\sin\left(\frac{m}{4}\vartheta\right)\right|^{n_3}}}.e^{k\vartheta} \qquad (9.21)$$

Generalized is understood in the framework of Pythagorean compact description. The transformation is a "curvature" related parameter, defining smaller or larger circles or spirals with particular shapes. In the case of spirals this transformation can also be applied as substitute for the parameter k, opening many other possibilities of studying natural phenomena. It is important to note that this can be completely embedded in higher dimensions, both in definite and indefinite setting, within the framework of contemporary geometry. When studying natural shapes, we will be able to recognize the major tendencies (circle or spiral, or constant ratio

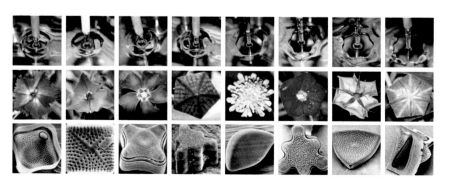

Fig. 9.8 Universal Natural Shapes

in general), and the local deformations induced by the specific circumstances via the curvatures. From a kinematic point of view, the movement traced out by the radius or position vector can be decomposed in a basic shape (circle or spiral) and a Gielis curve, in planar projection or as movements on (generalized) cylinders and cones or on supershapes (Fig. 5.9).

Determinatio and Constructio

In the *Determinatio* and *Constructio* parts it was initially shown how Lamé and Gielis curves are embedded in a long tradition of geometry and mathematics. They can be extended in several ways, such as varying spacing of angles, or combinations of shapes, while retaining their compact Pythagorean structure. Several methods can be applied in the study of natural shapes, from the solution of boundary value problems, to the geometric description of combined domains via R-functions. Also Fourier's method of separation of variables can now be applied to many problems in mathematical physics and to any normal polar domain in 2D and 3D, which will make life much easier.

We showed how with the description of the domains, various characteristics come for free, and how description itself can be used to define specific unit circles and as a measure of curvature. This is not restricted to 2 or 3 dimensions only. With the compact description a link was made to finite type polynomials, in particular Chebyshev polynomials. At the same time these wonderful polynomials have been generalized with Lamé-Gielis transformations. Applications of all these methods are manifold, in any direction of mathematics and the natural sciences. It combines ideas of many of the leading figures in the history of mathematics, and opens up Lamé's calculus for general applications in all fields of science, biology and physics. It turns out to be another proof of the beauty and magic of geometry. *All-is in Wonderland.*

Part V
Απόδειξις—Demonstratio

Chapter 10
Bamboo Leaves and Tree Rings

The plant has always appeared to us very remote, because its life is so still and silent. In its apparent immobility and placidity the plant stands in strong contrast to the energetic animal with its reflex movements and pulsating organs. Yet the same environment which, with its changing influences, so profoundly affects the animal, is also playing upon the plant. Storm and sunshine, the warmth of summer and the frost of winter, drought and rain – all these and many more come together and go about it. What about the subtle impressions they leave behind? Internal changes there must be; but our eyes have not the power to see them.

Sir Jagadis Chunder Bose

Shape Analysis with the Superformula

Our transformations are continuous and can capture a lot of variation in a few numbers. Assigning ranges to these few numbers should allow the capture of variation to some mathematically reasonable extent. These ranges can be determined from many repeated and reproducible measurements on real shapes. In this chapter we will see how variation fits well within the framework of generic and geometric description. This will be shown for bamboo leaves and for tree rings in various conifers or softwoods. Variation can be captured into a statistical distribution, and from these statistics the parameter ranges can be extracted. It will also be shown that for the study of nature, low parameter versions of our formula suffice. In fact, one parameter for dimension and one for shape are enough for describing bamboo leaves for 46 species studied, and also for the study of variation of tree rings.

Beyond the geometrical methods outlined in Determinatio and Constructio chapters, methods and algorithms to reconstruct Gielis curves and surfaces from data points and contours have been developed, both for 2D and 3D. These algorithms can detect supershapes such as starfish, quantifying the differences between various starfish. The algorithms can detect shapes with integer and non-integer

© Atlantis Press and the author(s) 2017
J. Gielis, *The Geometrical Beauty of Plants*,
DOI 10.2991/978-94-6239-151-2_10

symmetries, without prior knowledge, or training. It is important to note that the detection of rotational symmetries of real and incomplete objects without prior information is still a challenging and open problem. Furthermore, since symmetries are not differentiable parameters, they cannot be retrieved using deterministic methods, which imposes additional pre-processing and symmetry extractions before shape recovery [204, 205]. This problem has been solved, and the algorithms even allow the analysis of self-intersecting curves, which can be found for example in all projections in the plane of curves wound on helices or conics or shapes based on spherical products with self-intersecting cross sections. To our knowledge these are the first algorithms that can find self-intersecting curves, in the plane or in space. They are robust as well, against outliers or incomplete data [134].

The most efficient techniques are based on combinations of stochastic and deterministic methods. Evolutionary algorithms are used for a very limited number of generations to rapidly determine broad solutions that are then refined using deterministic methods. In the first step, the stochastic method provides one family of solutions for the raw problem of determining the appropriate symmetry and approximations of the shape coefficients without any prior information. Through a careful definition and construction of evolutionary operators coupled with a cost function built upon the Euclidean distance, the algorithm very quickly eliminates shapes with incorrect symmetries to only retain a population of few approximate solutions. Because a major drawback of stochastic methods is their slow convergence, the fine-tuning of an optimal solution is obtained in the second step using the Levenberg-Marquardt algorithm [134].

Consequently, this strategy allows for robust and efficient shape initialization with quick and accurate determination of a solution. By way of comparison: if only stochastic methods were used for the analysis, this would require thousands of generations for large populations, whereas now a few hundreds of generations for much smaller populations suffice. Oppositely, if only deterministic methods were used, some of the slightest variations of the initial guess, for instance due to missing data or incoherent pose estimation, might lead to local minima and/or numerical instabilities. Stochastic approaches need not be restricted to evolutionary algorithms, but other methods like particle swarm optimization *PSO* can be used to determine the optimal parameters for the optimal shape. This can even be improved by adding deterministic optimization in the final stages. These methods can be applied to abstract, man-made or natural shapes of highly complex nature. However, in nature we find numerous examples of much simpler cases in which only a limited set of the 6 parameters of the Gielis formula is needed, and we will provide detailed examples of bamboo leaves and tree rings.

Plant Leaves

Fundamental features of plants are self-similarity and symmetry, and all plants are in some sense repetitions of basic phytomers. Most plants, in particular monocots, have a simple structure, a repetition of phytomers consisting of an internode and a nodal zone. At the nodal zone a bud and protective leaf structures are implanted. This basic structure has led to a great diversity of growth forms and to a variety of plant models for plants, trees, grasses and bamboos, members of the grass subfamily Bambusoideae.

Plant growth is the result of the timing and spatial organization of the development of organs from a meristematic zone. In physiology various studies have focused on timing and spatial aspects of gene expression, in relation to the formation of organs. One example is the development of lateral outgrowth of organs from apical meristems and the direct relation to the spatial distribution of auxin via auxin transporters. However, gene action or gene regulatory networks do not give a complete picture. In the development of such organs it is mathematical and physical laws that are leading. It is no surprise that Fibonacci and Lucas series are found so frequently in phyllotaxy, both in the positioning of leaves along the stem and in the arrangement of parts in single and composite flowers. Phyllotactic patterns (Fig. 3.9) have also been linked to physical phenomena optimizing space utilization [206].

Such locally organized structures then develop into organs, into new phytomers or into leaves, either vegetative leaves or as modified leaves into floral structures. Vegetative leaves are the most important plant organs for photosynthesis, and their size and shape of leaves (Fig. 10.1) have an important influence on the morphogenesis and development of plants. As many plant organs vary with time during growing seasons, it is important to identify a simple but adequate mathematical model for capturing morphological change in plants. This is not a trivial task in plants. Besides the classic phenotypic differences related to ecological habitat, there are the phenomena of heteroblasty and heterochrony [207]. Heteroblasty is the phenomenon, whereby several types of leaves occur, depending on position, or age.

Fig. 10.1 A variety of plant leaves

Heterochrony on the other hand is the phenomenon whereby timing is crucial. Differences in timing can lead to simple versus compound leaves and heterochrony is at the basis for natural variation in *Cardamine hirsuta*, a close relative of *Arabidopsis*, whereby differences in leaf shape were correlated to flowering time. Last but not least it has been shown to be very difficult to define precise landmarks in leaves for morphometry in plants. Leaves and leaflets in compound leaves have in general relatively simple shapes, but leaves can be highly variable as in *Papaya* and *Vasconcellea* [208].

46 Bamboo Species, 1000 Leaves, 1 Equation

It started all and continues with bamboo. The bamboos (Poaceae: Bambusoideae) consist of about 1400 species of temperate and tropical woody bamboos and a tribe of herbaceous bamboos. Unlike other grasses, bamboos are the only major lineage within the family that have adapted to and diversified within the forest habitat. Bamboo leaves consists of two major parts, a sheath part and a blade part, with a characteristic ligule with oral setae and auricles as appendages. Cauline leaves have a protective function in the development of young shoots and stems. The sheath part is very stiff and the blade is reduced strongly. In contrast, foliage leaves have fully developed leaf blades and the blades are connected to the sheath via a pseudo-petiole. This is a structure unique to all bamboos and some large leaved grasses. These foliage leaf blades are usually linear, lanceolate or oblong-lanceolate, with the tip long acuminate, often scabrous and the side glabrous or softly hairy. The leaf blade is generally thinner than the culm sheath blade and often shows more marked dorsiventrality. The morphology and structure of leaves vary among species, and can be used for species identification [209] (Fig. 10.2).

In contrast to all other grasses (with the exception of some large leaved grasses) the leaf blades of bamboo are connected to the sheath part by a pseudo-petiole, typical of all bamboo foliage leaf blades, reflecting the evolutionary stability of this solution. Indeed, bamboos are one of the very few groups in the grass family that have evolved in forests, not in open areas, still with large sizes (up to 20 m for temperate bamboos and up to 30 m for tropical bamboos). Hollow and flexible culms and branches, protected during development with very stiff culms or branch sheaths on the one hand, and leaf blades that are connected through pseudo-petioles allowing for torsion on the other, contribute to the evolutionary success of woody and herbaceous bamboos.

Also the anatomy of bamboo leaves is very stable throughout the subfamily. It is remarkable that the pseudopetiole, a specialized structure in all bamboos (and only some grasses with larger leaves), has hardly been studied. The course of the vascular bundle and the possible presence of pulvini should be investigated, since pseudopetioles allow bamboos to adjust the orientation of the blades towards the sun or to ease torsion caused by wind or snow loads.

Fig. 10.2 *Sasa palmata*

A Two-Parameter Model

The Gielis formula was tested on 46 bamboo species from different genera (Table 10.1). For each species, the length was measured as the distance from the leaf bottom to the tip, for over 500 leaves. 30 leaves per species of (approximately) equal length close to the species mean were then selected and scanned for extracting the coordinates on the leaf boundary.

Table 10.1 46 Species of bamboo [210]

Code	Latin name	Genus
1	*Bambusa emeiensis* var. *viridiflavus* Hsuen et Yi	*Bambusa*
2	*Bambusa multiplex* (Loureiro) Raeuschel ex Schultes and J.H. Schultes	*Bambusa*
3	*Bambusa multiplex* f. *fernleaf* (R.A. Young) T.P. Yi	*Bambusa*
4	*Bambusa multiplex* var. *riviereorum* Maire	*Bambusa*
5	*Chimonobambusa marmorea* f. *variegata* (Mitford) Makino	*Chimonobambusa*
6	*Chimonobambusa neopurpurea* Yi	*Chimonobambusa*
7	*Chimonobambusa quadrangularis* (Franceschi) Makino	*Chimonobambusa*
8	*Chimonobambusa sichuanensis* (T.P. Yi) T.H. Wen	*Chimonobambusa*
9	*Chimonobambusa tumidissinoda* Hsueh and YI, ex. Ohrnb	*Chimonobambusa*
10	*Indosasa shibataeoides* McClure	*Indosasa*
11	*Oligostachyum sulcatum* Z.P. Wang and G.H. Ye	*Oligostachyum*
12	*Phyllostachys arcana* cv. luteosulcata McClure,	*Phyllostachys*
13	*Phyllostachys aurea* Carrière ex Rivière and C. Rivière	*Phyllostachys*
14	*Phyllostachys aureosulcata* McClure	*Phyllostachys*
15	*Phyllostachys aureosulcata* f. pekinensis J.L. Lu	*Phyllostachys*
16	*Phyllostachys aureosulcata* f. spectabilis C.D. Chu and C.S. Chao	*Phyllostachys*
17	*Phyllostachys bissetii* McClure	*Phyllostachys*
18	*Phyllostachys dulcis* McClure	*Phyllostachys*
19	*Phyllostachys edulis* (Carrière) J. Houzeau	*Phyllostachys*
20	*Phyllostachys glauca* McClure	*Phyllostachys*
21	*Phyllostachys heteroclada* Oliver	*Phyllostachys*
22	*Phyllostachys edulis* 'Gracilis' (W.Y. Hsiung C.S. Chao) and S. Renvoize	*Phyllostachys*
23	*Phyllostachys nidularia* Munro	*Phyllostachys*
24	*Phyllostachys nigra* f. *henonis* (Mitford) Muroi	*Phyllostachys*
25	*Phyllostachys nigra* (Loddiges ex Lindley) Munro	*Phyllostachys*
26	*Phyllostachys sulphurea* var. *viridis* (Carrière) Rivière and C. Rivière	*Phyllostachys*
27	*Phyllostachys violascens* (Carrière) Rivière and C. Rivière	*Phyllostachys*
28	*Pleioblastus argenteostriatus* (Regel) Nakai	*Pleioblastus*
29	*Pleioblastus chino* (Franchet and Savatier) Makino	*Pleioblastus*
30	*Pleioblastus distichus* (Mitford) Nakai	*Pleioblastus*
31	*Pleioblastus fortunei* (Van Houtte) Nakai	*Pleioblastus*
32	*Pleioblastus gramineus* f. *monstrispiralis* (Y. Okada) Muroi and H. Hamada	*Pleioblastus*
33	*Pleioblastus kongosanensis* f. *aureostriatus Muroi and Y. Tanake*	*Pleioblastus*
34	*Pleioblastus maculatus* (McClure) C.D. Chu and C.S. Chao	*Pleioblastus*

(continued)

Table 10.1 (continued)

Code	Latin name	Genus
35	*Pleioblastus simonii* f. *heterophyllus* (Makino and Shirasawa) Muroi	*Pleioblastus*
36	*Pleioblastus yixingensis* S.L. Chen and S.Y. Chen	*Pleioblastus*
37	*Pseudosasa amabilis* var. *convexa* Z.P. Wang and G.H. Ye	*Pseudosasa*
38	*Pseudosasa japonica* var. tzutsumiana (Siebold and Zuccarini ex Steudel) Makino ex Nakai	*Pseudosasa*
39	*Semiarundinaria densiflora* (Rendle) T.H. Wen	*Semiarundinaria*
40	*Semiarundinaria sinica* T.H. Wen	*Semiarundinaria*
41	*Shibataea chinensis* Nakai	*Shibataea*
42	*Sinobambusa tootsik* (Makino) Makino	*Sinobambusa*
43	*Indocalamus pedalis* (Keng) P.C. Keng	*Indocalamus*
44	*Indocalamus pumilus* Q.H. Dai and C.F. Keng	*Indocalamus*
45	*Indocalamus barbatus* McClure	*Indocalamus*
46	*Indocalamus victorialis* P.C. Keng	*Indocalamus*

The shape of all leaf bamboo species is described by the simplified Gielis equation extremely well [209, 210]. The shape of bamboo leaves is encoded in two parameters, one for shape n and one for size (Eq. 10.1; Fig. 10.3), and the predicted leaf shape matched the observed leaf shape perfectly for bamboo leaves (Fig. 10.4).

$$\varrho(\vartheta) = \frac{l}{\sqrt[n]{\left|\cos\left(\frac{\vartheta}{4}\right)\right| + \left|\sin\left(\frac{\vartheta}{4}\right)\right|}} \tag{10.1}$$

Fig. 10.3 Leaf shape and size parameters n, L, l are related $L = \left(1 + 2^{-1/2n}\right) \cdot l$. Copyright Nanjing Forestry University, Bamboo group

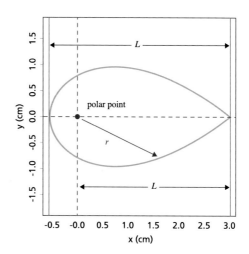

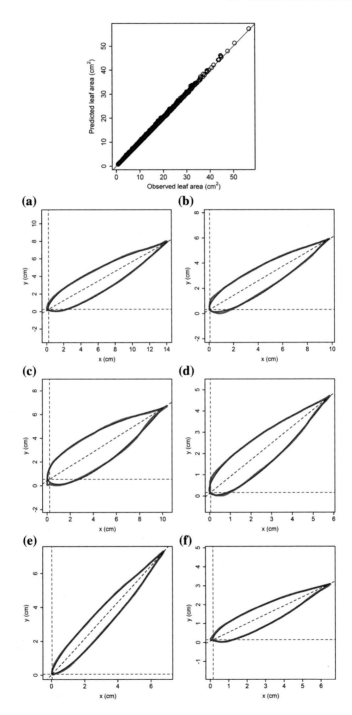

Fig. 10.4 Comparison between scanned leaf profile and predicted leaf profile from the simplified Gielis equation for **a** *Pleioblastus yixingensis*; **b** *Bambusa emeiensis var. viridiflavus*; **c** *Indosasa shibataeoides*; **d** *Phyllostachys bissetii*; **e** *Phyllostachys edulis*; **f** *Phyllostachys edulis "Gracilis"*. Copyright Nanjing Forestry University, Bamboo group

The goodness of fit demonstrated also provides convincing evidence of bilateral symmetry in bamboo leaves, with all coefficients of determination higher than 0.980; the regression line only trivially deviated from the straight line of $y = x$. Fig. 10.4

The variation of size and shape-parameters for these 46 species is illustrated in Figs. 10.5 and 10.6. Although there was significant difference in the leaf-shape parameter among species from different genera or from the same genus, the calculated leaf-shape parameters for these 46 species range only from 0.02 to 0.1. The lower values are for shapes of more linear-lanceolate type, with *Pleioblastus chino, Pleioblastus simonii* f. heterophyllus, *Pleioblastus gramineus* f. *monstrispiralis, Chimonobambusa tumidissinoda* and *Phyllostachys edulis*. Leaves described by the higher values of the shape parameter n are slightly broader in shape, with *Shibataea chinensis, Indosasa shibaeatoides* and *Bambusa multiplex* var. riviereorum. The rest of the leaves fall broadly within the range $n = 0.03$–0.08. The leaf shape is only slightly variable in *Indocalamus* species.

Modelling with the modified Gielis equation with reduced number of parameters is clearly applicable to a wide variety of temperate bamboo genera and species, with lanceolate leaves. The range of shapes and sizes, from the small leaves of *Bambusa multiplex* var. riviereorum to large leaves of *Indocalamus victorialis*, and from the linear-lanceolate leaves of *P. linearis* to the broader leaves of *Shibataea*, can efficiently be modeled with one equation, with only two parameters.

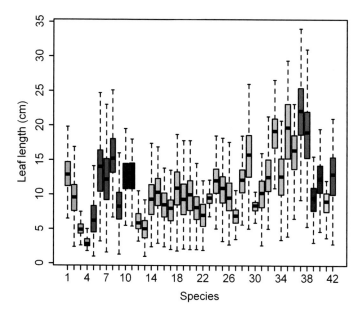

Fig. 10.5 Leaf lengths of 42 species for different genera; numbering in Table 10.1. Copyright Nanjing Forestry University, Bamboo group

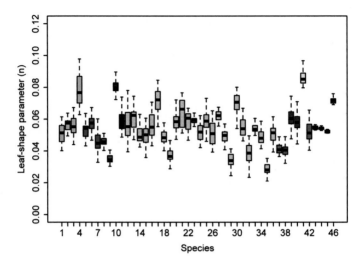

Fig. 10.6 Leaf shape parameter *n* for 46 species of bamboo. Color codes are used for different genera; numbering in Table 10.1. Copyright Nanjing Forestry University, Bamboo group

The length of all leaves of these bamboo species follows the Weibull distribution, rather than the normal distribution. This distribution originates from the study of different particle sizes in crushed particles, sand or volcanic ash, or failure rates in products, following power laws. This implies that the distribution of leaf length in bamboo follows a power law. Although the leaf sizes (length and width) of the 46 bamboo species of the study differ considerably, their leaf shapes vary relatively little, but the difference can be clearly identified. The shape of bamboo leaf blades, connected to a pseudopetiole, is a very efficient solution in evolutionary sense. This solution of a large leaf connected via a pseudopetiole, with leaf blades of varying size to have both protective and light harvesting functionality, has been used in (almost) every bamboo, both on macroscopic and anatomical level [211].

Whereas in the past bamboo leaf blades were characterized by the quantitative measures of length and width with additional qualitative characteristic such as linear-lanceolate or oblong-lanceolate, we now have a clear quantitative number for a qualitative characteristic. Besides the use in identification of species, this can also be employed as a very precise method in ecology, taxonomy and studies of genetic diversity of bamboo, avoiding the use of molecular markers. Only two parameters are involved, one for length and one for shape, accounting for full variations in shape and dimensions compared to 3 dimensional and 3 shape parameters for grasses [209].

Superelliptical Tree Rings and Bamboo Stems and Rhizomes

Gielis transformations were used to study tree rings of ten different conifers or softwoods. In general, cross sections of trees hint at circular shape of tree rings. Tree rings that are more square or deviate otherwise from circles are found in sections of square stems or branches (as is the case in teak branches and young stems of teak), or sections close to roots (e.g. square oak, anchoring with four roots) or higher up the stem in trees with butt roots (Fig. 10.7).

It was shown in [212] that tree rings are much better modeled by superellipses (Fig. 10.8). The modified version of the Gielis formula used is with all exponents n equal, which are Lamé curves, and with a and b, which starts from superellipses (Eq. 10.2). Measuring hundreds of tree rings proved that the model was very efficient.

$$\varrho(\vartheta) = \frac{1}{\sqrt[n]{\left|\frac{1}{a}\cos\left(\frac{\vartheta}{4}\right)\right|^n + \left|\frac{1}{b}\sin\left(\frac{\vartheta}{4}\right)\right|^n}} \quad (10.2)$$

In this study this model was compared to a classical model of tree rings regarded as circles using the Aikake Information Criterion. This is one statistical method, which is a trade-off between the model's success in fitting the data and the complexity of the model; it gives a relative measure of how efficient and simple a model is. If two models have the same success in data fitting, the better one in the AIC sense is the one with fewer parameters. A model with two parameters for describing

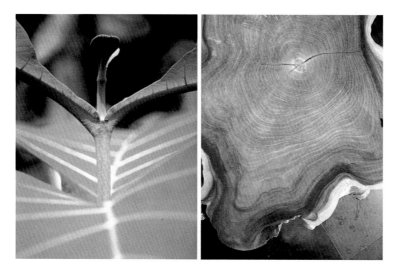

Fig. 10.7 Young stems of teak and cross section of teak stem

(a)   **(b)**

(c)   **(d)**

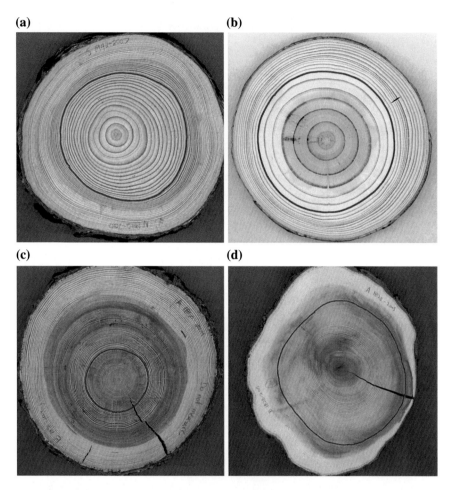

Fig. 10.8 Superelliptical tree rings in softwoods. Clockwise from *upper left* Jack pine (*Pinus banksiana* Lamb), red pine (*Pinus resinosa* Aiton), tamarack (*Larix laricinia* (Du Roi) K. Koch and white cedar (*Thuja occidentalis* L.). Copyright Nanjing Forestry University, Bamboo group

bamboo leaves in two parameters is preferred to a static model of grass leaves with six parameters.

Both in bamboo leaves and tree rings there are excellent agreements between model and real measured data, but in both cases two parameters suffice. So definitely from the AIC point of view the model is preferred to other multiparameter models in the case of grass leaves, or to circular tree rings. In the latter case the AIC of circle and superellipse are only the same when the tree ring is really circular. In the case of the superellipses, the AIC remains low and stable across a range of exponent n (from 0.75 to about 3) using simplified versions (Lamé type) [212].

The study of tree rings permits a quantification of the rotation of tree rings over the years, and it is one of the very first in its kind to focus on this aspect of growth in trees. In softwoods it is well known that the stems of softwoods tend to spiral, and may change direction after some time.

This rotation around the vertical axis is also torsion, relative to the axis as a cylinder, and it has both genetic and environmental components. It is one combination of vertical and spiral tendencies, a field of study, which inspired not only Goethe's *Über die Spiraltendenz der Vegetation* (1831) and Darwin's *On the movement of climbing plants* (1875) [213], but also Walter Liese's detailed *Anatomische Untersuchungen an extrem drehwüchsigem Kiefernholz* [214]. There is a lot to be discovered and understood in this exciting field of research in plants and wood, where we combine our new geometrical methods with the excellent work of excellent predecessors (Fig. 10.9).

A recent study by the Nanjing group on bamboo shoots and rhizomes used our equation to study the shape of culms and pith cavities in various bamboo species [215]. Stems, shape of shoot apical meristems and the spirality in certain bamboos (e.g. monstrispiralis) can efficiently be modeled. The original idea of using Lamé curves to study square bamboos, *Chimonobambusa quadrangularis* (number 2 in the series in Fig. 10.10) has extended to modeling all bamboos, stems, rhizomes, meristems, tissues…. In Fig. 10.10 numbers from 1 to 8 represent eight bamboo species, respectively: *Phyllostachys bissetii, Chimonobambusa quadrangularis, Phyllostachys makinoi, Phyllostachys edulis, Bambusa polymorpha, Dendrocalamus barbatus* 'internodiradicatus', *Dendrocalamus latiflorus* and *Dendrocalamus sinicus*. Both wall and pith rings of all species can be very efficiently modeled with our transformations, and this works for sections along stems and rhizomes, including the spirality as in *Pleioblastus* Monstrispiralis, for cross sections of meristems as well, as shown in [215].

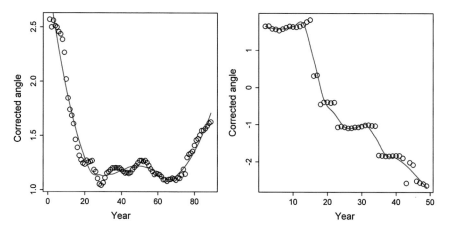

Fig. 10.9 Quantification of spirality in softwood trees. Copyright Nanjing Forestry University, Bamboo group

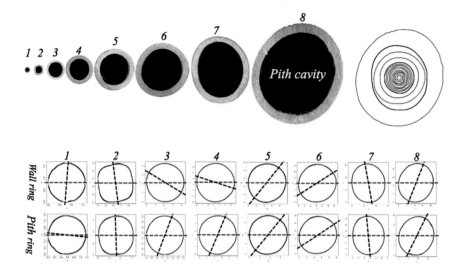

Fig. 10.10 Observed culm wall rings and pith cavity rings (*black solid lines*) and the predicted rings (*red solid lines*) of eight bamboo species show excellent correspondence [215]. Copyright Nanjing Forestry University, Bamboo group

Stems, Petioles and Cells

With our transformations we can initiate the geometrical study of these tendencies in nature. Having one formula permits for the exact calculation of associated characteristics such as perimeter, area and polar moment of inertia, with a definite advantage over classical approaches using engineering formula. This provides a direct advantage for biologists dealing with variable shapes and biomimicry. The associated characteristics come for free with the description and actual shapes result from solutions to classical boundary value problems.

In a study on the structural morphology of petioles of *Philodendron melinonii* (Fig. 10.11) and the various layers of different tissues the shape of the petiole was modelled with the Gielis formula [216]. Previously researchers had to resort to specific formulas from engineering, for beams with well-defined cross sections (elliptic, square, circular, T or I beams…). Now shapes can be studied with well defined formulas over a continuous change of cross-sectional geometry along its length and layer thickness, and to model different layers with different mechanical characteristics. This allowed for gaining insight into the efficiency of biological beams described by different tissues properties, geometry and turgidity, with *Philodendron* and rhubarb petioles as object of study, combining theory and measurements [216, 217]. The sections of petioles were scanned and analysed using the full six parameters set of the Gielis transformations (Fig. 10.12).

In the same spirit one can efficiently analyse square stems or columnar cacti, using subsequent cross sections at different heights. In the genus *Chimonobambusa*

Fig. 10.11 *Philodendron melinonii* with green petiole and *P. melinonii* "Rubescens" with red petioles

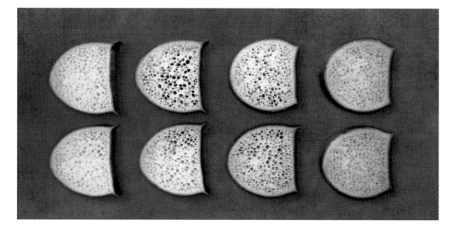

Fig. 10.12 Series of sections of petioles of *Philodendron melinonii*

for example, *Ch. quadrangularis* is square-like along its complete stems, but other species of the genus have square stems only in the lower zones of the stems. Also cross sections of underground stems of Moso bamboo have been modelled in this way. The model is also dynamic: the volume and size and shape of cross sections (area and perimeter) of cacti changes depending on how much water is stored in the cactus: it will expand after rainfall, shrink in drier periods

Many plants have square stems. In *Inventing the Circle*, the example was given of a square stem with the same base radius as a circle. Growing the Lamé super-circle from exponent $n = 2$ to $n = 5$ resulted in extra area in the corners of about 21%, but the polar moment of inertia increased by more than 50%. A small investment in extra area and material then results in extra strength. The area/polar moment of inertia ratio is but one example, and focused only on homogeneous materials, but plants are smart: they invest in extra strong materials and tissues in these corners and the weakest tissues in the centre (or even no materials at all-in hollow stems) for the same geometrical reason: the tissues farthest away from the central axis have the most prominent role in resistance against torsion and bending [1, 215, 216, 218].

Having these characteristics available immediately allows to study such ratios in more depth. Ratio's as area/perimeter, area/polar moment of inertia etc. can be used. One can compare this to the mathematical geometrical question of the isoperimetric inequality: what shape has the largest area for a given perimeter? We know since Antiquity that this is the circle, but if one asked what is the best shape for the optimal ratio of area/perimeter, circle and square both win, with ratio equal to 2 for both. With other constraints, we will find supershapes with exponents different from 2 or 1 to be the better solutions [218, 219]. These methods blend in seamlessly with current methods used in plant biology, biomechanics [220] and biophysics [221] (Fig. 10.13).

Instead of studying cross sections along a stem or petiole, we can use the very same methods to understand positioning of shapes in a plane, on a curved surface or in space. Similar optimisation principles are at work in tissues. Cells in plants are stacked in ways to optimize the internal volume, minimize the intercellular space and maximize the area of exchange between cells, while ensuring strength. Wood is one example, but Nature has provided many, many examples of beautiful

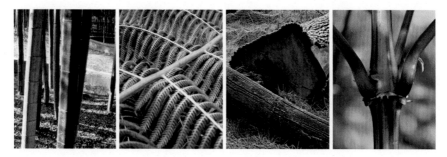

Fig. 10.13 Stems of bamboo, petioles of tree ferns, the square basis of an oak and the stem of *Impatiens glandulifera*

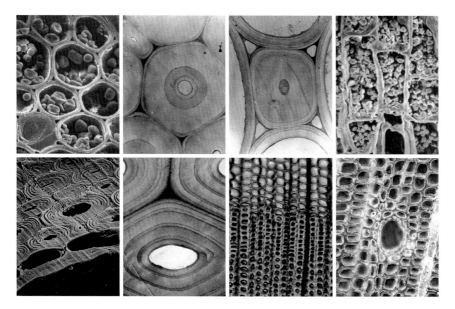

Fig. 10.14 Wood anatomy: *Upper row* bamboo parenchyma with starch (*left and right*) and bamboo fibres. (*center left and center right*). *Lower row* fibres of rattan (*left and left center*) and softwoods (*right center and right*). Copyright Walter Liese, Marc Stevens and Gudrun Weiner

multi-optimization problems, based on purely geometrical strategies. Improved modeling accuracy will have profound effects on estimating biomass, deriving more accurate allometric laws and so much more (Fig. 10.14).

The many applications of our transformations in technology, e.g. nanotechnology, optics, antenna and wireless communications [222, 223] could lead to a better understanding of nature. Alternatively, natural shapes could serves as an endless source of inspiration for the development of technology, e.g. *Ginkgo* leaves for wearable antennas [224]. We now study these problems based on one single unifying transformation, indifferent of the scale or the nature of the problem *(independent of the forces that create them)*. The exact mechanisms underlying this optimization in nature can be of physical, chemical or biochemical nature. To adjust to the environment in an optimal way organisms and natural objects can draw on chemistry, genes, from the existing toolbox. One thing is certain: the systems we study are structurally stable. The kosmos becomes more cosmos or ordened, rather than chaos and chaotic.

Chapter 11
Snowflakes and Asclepiads

There is grandeur in this view of life, with its several powers,
having been originally breathed into a few forms or into one;
and that, whilst this planet has gone cycling on according to the
fixed law of gravity, from so simple a beginning endless forms
most beautiful and most wonderful have been, and are being,
evolved.

Charles Darwin

Soap Bubbles, *Culcita Novaeguineae* and Ammonoids

One of the most beautiful stories in botany is that of the Orchid *Angraecum ses-quipedale*. It has very long nectar trails and Charles Darwin inferred the existence of a certain species of pollinator, having long tongues [225]. Alfred Wallace provided guidance towards the discovery of this moth [226]. 50 years later such animal was discovered and it was aptly called praedicta as its existence had indeed been predicted. If one looks at the spherical shape of a free floating soap bubble, the sphere seems to be the best shape to minimize tension. Soap molecules can organize themselves in all directions to achieve this. In crystals, flowers and starfish, however, there are preferred directions; not all parts can move freely in the same direction to achieve the shape of a sphere. On the contrary: in nature we see square stems and starfish as optimized solutions, following millions of years of adaptation to natural conditions and environments. Square stems represent optimization of fourfold symmetry, starfish of five fold symmetry with a lot of variation on these basic themes.

A natural question is then: if we have soap films on the one hand and starfish on the other, nature should have experimented with intermediate solutions as well, shapes somewhere halfway between sphere and starfish shape. If one asks the question, the answer is already given: searching through collections of starfish, I soon came across *Culcita novaeguineae*, a genus of starfish from tropical waters in Indian and Pacific Ocean. These starfish are also known as cushion stars, and are "pentagonal spheres". Interestingly the juvenile form of these starfish is a flat pentagon, inflating into a cushion star (Fig. 11.1).

© Atlantis Press and the author(s) 2017
J. Gielis, *The Geometrical Beauty of Plants*,
DOI 10.2991/978-94-6239-151-2_11

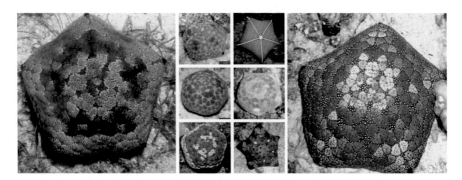

Fig. 11.1 *Culcita novaeguineae* var. praedicta. Copyright Poppe Images

In the same way we can study diatoms in their own geometry. A prepared mind equipped with the new glasses can discover in natural sciences and natural history an endless number of examples that can be described with our transformations. The more one looks, the more we find. In 2012, a superelliptical galaxy was discovered [227], there are square bacteria in salt conditions [228], one finds hexagonal patterns on Jupiter, or pear shaped nuclei in atoms [229], easily modelled as one-angles in the same way bird's eggs are one-angles [1]. Universal Natural Shapes refer to natural shapes at all possible levels: in growing organisms, fluid mechanics, snowflakes, galaxies and elementary particles.

Singling out one example: triangular and square coilings of certain ammonite fossils, like *Entogonites saharensis* (Fig. 11.2) and *Soliclymenia paradoxa* are well known to experts. These are not some strange experiments, peculiarities, degeneratives or freaks of nature, but these genera thrived in their specific environments, even with a wide geographical distribution, from the Late Devonian up to 360 Million years ago [230–232].

In Fig. 5.16 Future Fossils we encountered some very peculiar specimens harvested from our supershaped universe. One of those had some peculiar involutions. Actually, it resembles the structures of certain fossils of the ammonoids *Wocklumeria* and *Parawocklumeria* that also show this triangular symmetry with

Fig. 11.2 *Left* and *center left* fossil and reconstruction of living *Entogonites saharensis*. *Center right Parawocklumeria. Right* a discovery from the supershape universe. Copyright Dieter Korn

Fig. 11.3   Plant pollen. Copyright National Botanic Garden Belgium

involutions, although less pronounced then our Future Fossil. This fits well within a morphological (and proposed phylogenetic) series of this group of ammonoids.

Equilibrium Shapes for Far-from-Equilibrium Conditions

Such shapes (Fig. 11.3) are results of optimization problems, much in the same way as soap bubbles and films are solutions to optimization problems, mathematically described as constant mean curvature surfaces. The extension of spheres to Wulff shapes and supershapes as unit spheres has been discussed before, but one should also take into account differences in solutions in different directions. A cell in a growing stem of plants will keep its width, but expand vertically. D'Arcy Thompson wrote the equation as a generalized Laplace-Young equation taking into account the curvatures, but with different tensions T_1 and T_2 in perpendicular directions. Instead of the arithmetic mean of the curvatures we have a weighted arithmetic mean. Koiso and Palmer [13] interpret this in the framework of constant anisotropic mean curvature surface CAMC:

$$\Lambda = \frac{T_1}{R_1} + \frac{T_2}{R_2} = constant \qquad (11.1)$$

where Λ is the anisotropic mean curvature, $1/R_1 = \kappa_1$ and $1/R_2 = \kappa_2$ the principal curvatures of a smooth surface, and T_1 and T_2 are orthogonally directed tensions which depend on the material and the normal directions of the surface at each point. Λ constant means that the anisotropic mean curvature is constant on the whole surface and this surface then is called a CAMC surface.

Natural snowflakes provide nice examples. "Triangular" snowflakes retain their hexagonal symmetry as supershapes, with alternating short and long sides, thereby reflecting the molecular symmetry of ice. The first systematic study on snowflakes goes back to Kepler's *Strena seu de nive sexangula* (1611), in which the six-cornered snowflake was interpreted for the first time in terms of the underlying symmetry. Kepler based his consideration on close packing of spheres, and later the underlying molecular symmetry of ice turned out to be six. The geometrical description of snowflakes has not seen great progress, with the exception perhaps of Aloysio Janner's *De nive sexangula stellata* [233, 234].

In the past century quite a lot of research has been performed to understand the physical basis. With a fixed molecular symmetry, the growth and actual shape of the snow crystal is dependent on a range of parameters. Snowflake diversity is enormous, and one finds, among others, dendritic types, plates, (hollow) columns and capped columns. From a physico-mathematical perspective, little progress has been made. A major open problem in crystal formation is the existence of equilibrium shapes.

Equilibrium surfaces have not been observed for snowflakes, at least not in the classical sense and surface energies are thought to play negligible roles in snowflake formation [235]. Spheres or classical Delaunay shapes, as equilibrium shapes, have never been observed in snowflakes. When anisotropic energies are considered however, a range of shapes of snowflakes can be seen as constant anisotropic mean curvature CAMC surfaces for anisotropic energies. Such shapes then are equilibrium shapes for processes where very short formation times and long range correlations (typical of far-from-equilibrium thermodynamics) dictate the development.

The existence of equilibrium shapes described by CAMC surfaces in crystallography has the following results: First, a unifying descriptive approach for crystal formation with very different formation times; second, it suggests strongly that surface energies do play an important role, also far-from-equilibrium, and finally, (snow)crystal growth can be embedded into the broader framework of submanifold theory, the theory of shape. It is mentioned in passing that a wide range of the shapes encountered in crystallography (cubes, prisms, triclinic, hexagonal...) are simply supershapes. The shapes can easily be turned into hexagons, by admitting one of the cross sections to become hexagons instead of squares. The resulting Wulff shape will be a hexagonal prism, while the resulting minimal surface will be a supercatenoid with hexagonal cross section, planes and (almost) right angles between planes and caps (Fig. 11.4) [236].

Capped columns, simple prisms, needles, cups, bullets and other shapes of snowflakes can thus be modeled and understood as equilibrium shapes with anisotropy already from the very beginning, and this anisotropy is retained in fully developed snowflakes. It should be clear that actual snowflakes are the result of very complicated physical processes, but that CAMC surfaces as equilibrium shapes for far-from-equilibrium conditions, provide a novel way of understanding

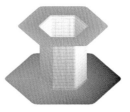

Fig. 11.4 Capped column snowflake and CAMC catenoid

the aforementioned major open problem in snowflakes, that of equilibrium shapes and surface energies.

Supershapes provide equilibrium shapes at all levels in nature (static or dynamic). It seems in fact quite straightforward to look at flowers, sea lilies and sand dollars. During growth, development and during lifetime in general, they try to minimize certain anisotropic energies. In the same way as the shapes of snowflakes are the result of the molecular symmetry of ice which is imposed from the early stage of nucleation and retained throughout, the symmetry of flowers is the result of the earliest stage of development, with the symmetry imposed from the start by the flower buds. Petals are solutions to minimize tension and stress. The shape and development of petals (as one part of the flower) is a mechanism to get rid of stress. If a pentagonal cap is placed upon a developing square meristem, then it would grow into a flower with five petals and sepals, not four; it is a minimization or optimization to reduce stress, in the very directions and shapes indicated by the curvatures.

In flowers, the development of floral organs like sepals and petals is related to getting rid of stress by expanding relatively planar structures. From a small meristem of tenths of milimeters, they develop into smaller or larger structures (in some cases up to flowers of one meter or more). Stresses might be internalized in other ways as well. The curvatures on a sand dollar like pentagon fit well within the pentagonal shape. In this case the stresses are internalized in a growing but essentially static structure, but are then employed as functional entities: the floral structure on the tests of sand dollar (Fig. 11.5) are actually used as opening from

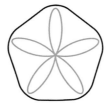

Fig. 11.5 Sand dollars and curvature

which the tube feet extend [237]. Development of natural organisms is a beautiful symphony, for those who are open to its sounds.

Fusion as Driving Force in the Evolution of Angiosperms

The development of flowers is a major evolutionary step in plants, leading to gymnosperms and angiosperms, and the major subdivisions of the latter into monocots and dicots, with a variety of pollination syndromes. In particular, flowers enabled the evolution of insect-mediated pollination, a driving force in the rapid evolution and radiation of both flowering plants and pollinators. In the plant world there are a large number of highly fascinating plants, with the most ingenious methods of survival, pollinator deception, seed dispersal and much more. Few are as fascinating as orchids. It was Darwin himself who started in depth research on the fantastic pollination syndromes of orchids with many ingenious ways to attract (and deceive) pollinators [238]. The attractive clues can be of visual or olfactory nature, with the lip of the orchid resembling a female insect body, or the olfactory clues matching precisely the olfactory composition of pheromones of female insects.

One peculiar aspect of the orchid family is that the pollen is glued together in pollinia, and it is the pollinia that are attached to the deceived pollinator, flying off to another flower to reap real rewards (to be deceived once more). Aggregation of pollen into pollinia in Orchidaceae proved to be highly efficient in evolution, increasing success in fertilisation. Innovations to increase the efficiency of pollination combine pollinia with unique structures for attachment to pollinators that often reduce self-pollination. In orchids such accessory structures are produced in whole or in part by the stigma, so that pollen units have origins in both male and female units. This is due to a special structure in which female (gynoecium) and male (androecium) organs are fused into the orchid column. This is one of the main reasons for the evolutionary success of orchids and the species richness of the family [239, 240].

Our knowledge of flowers, their diversity, development and rapid evolution is still very limited. A general model, almost three decades old now is the *ABC* model [241, 242]. Three groups of genes define four different whorls in flowers, with spatial restriction of the action of the genes, in sepals (only *A*), petals (*AB*), stamens (*BC*) and pistils (only *C* genes active). The *ABC* model originated from studies in *Arabidopsis* and *Antirrhinum*, whereby the various functions were identified via loss of expression of the various groups, or via expression of well-identified genes in other species. The model is basically simple, but less ubiquitous than it was originally intended to be [243–245]. For example, the *A* function has been shown to be less conserved and its function less clear, if any at all, in other plant species.

Inter and Intrawhorl Fusion in Flowers

Even considering some general applicability of the basics of the *(A)BC* model we are only at the very beginning of understanding development and evolution of flowers. Improved understanding of the flower as a determining factor in the evolution of angiosperms, should involve the study of specific trends in the evolution of floral morphology. Like the *(A)BC* model, such trends are an abstraction of nature, but they have the advantage to result from observations spanning the whole group of flowering plants and their ancestors, not only model plants.

One major trend in the evolution of the flower is the reduction of polymery to oligomery, from numerous to fewer [246]. Polymery is associated with spiral phyllotaxy, an undifferentiated perianth, the presence of staminodes, complex floral vasculature, and typically trimery or dimery as the basic merosity of the flower. Polymery is observed in *Nymphaea* and *Magnolia*, or many monocot flowers (Fig. 11.6).

Oligomery is associated with whorled phyllotaxy and a basic condition of diplostemony, whereby the number of stamens is twice the number of petals; the outer whorl of the stamens is oriented in alignment with the sepals, the inner whorl with the petals [247].

A second major trend is synorganization, in which two or more specific organs or structural elements undergo fusion, within and between whorls, thus enabling the conception of new functional entities. A whorled arrangement, radial symmetry, and a small, fixed number of floral organs (oligomery) are the major prerequisites

Fig. 11.6 Spiral phyllotaxy (*upper row*) and whorled phyllotaxy

Fig. 11.7 *Petunia hybrida, Petunia* Maewest, Wild type WT and Pink Ice. Copyright Michiel Van den Bussche

for the evolution of complex synorganizations. Synorganized structures are more stable and enable the formation of larger flowers, opening up new ways to experiment with shape and function, such as the formation of three-dimensional corollas with increased petal surface, enabling a wide spectrum of pollination syndromes [246, 248, 249].

Inevitably the genetics of fusion will be unraveled but at present, to our knowledge, the only molecularly defined mutant that exhibits lack of fusion of both carpels and petals is the *Petunia maewest* mutant [250, 251]. In *Arabidopsis* fusion of sepals and petals is observed only in a specific mutant. For fusion to occur growing organs must come into contact early during development and both size and spatial separation of individual organs are important. The prevalence of fusion in the perianth in dicots is associated with pentamery as the basic merosity, whereas in tetramerous flowers of *Arabidopsis* and in *P. maewest* mutants, the petals are spaced too far apart to initiate fusion [252] (Fig. 11.7).

Orchids and Asclepiads

The same is true in monocots, where tepals are spaced 120° apart, hence the separate tepals and lips in orchids. Despite the lack of fusion in sepal or petal whorls, the fusion events between male and female whorls have largely driven the evolution of orchids. Orchids are monocots but within the dicot group there is an equally interesting group of plants, with absolutely fascinating ways of attracting pollinators. These are the Asclepiads. They are small succulents growing in semi-arid conditions in Asia and Africa, and they have evolved olfactory systems resembling putrefaction, to attract insects preying on dying animals [253–255]. The flowers however are absolutely splendid, among the most beautiful in the plant kingdom, with highly varied shapes and colors, and a highly complex pollination apparatus with putrid or carrionlike odours. Fascinating colors in connection with special structures fostering deceit, trapping and attachment of pollinia: the

Fig. 11.8 *Tridentia gemmiflora.* Copyright Martin Heigan

Asclepiadaceae are known as the orchids among the dicots, with flower sizes ranging from a few millimeters to forty centimeters (Fig. 11.8).

In contrast to orchids, the pentamerous symmetry has allowed for fusion of petals, leading to a large corolla. In Stapeliads (*Asclepiadoideae*) it is the thick and fleshy corolla that accounts for this size variation, not the inner organs. The corolla originates from fusion of petals and forms a very stable structure necessary for the development of the highly complex pollination apparatus located in the center. This apparatus results from a very complicated and highly coordinated sequence of fusion events, following the initiation of the different separate whorls in the flower. Fusion occurs immediately following formation of the structure (congenital), or after the formation has been finalized (postgenital). Following their formation petals start to fuse and also later free parts are fused, and the fused petals are partly released at anthesis. The congenital union of stamens is followed by fusion of stamens with petals into a *corona*. This is shown schematically, following [255], in Fig. 11.9.

Postgenital fusion of carpels leads to the formation of a style head, to which stamens are fused forming a *gynostegium*. This then leads to the formation of guide rails and pollinaria. The pollinia are formed in the stamens, and the translator is secreted at the surface of the style head. The firm connection between anthers and style head by postgenital fusion is necessary for the formation of the *pollinarium*. This complex organ, anthers plus style head is called *gynostegium*. Following the independent formation of the (apocarpous) carpels, the apices fuse into a style head, then later fuse with anthers. The flower is a very complex pentamerous flower, with complicated internal chambers. The fertile parts are hidden in these chambers (internation of the fertile parts) [255].

The absolutely remarkable fact is the extreme stability in this group of plants. The mutual positions of the inner organs (part of the corona, stamens, carpels) are strongly fixed and precise. The size of the flowers is very diverse, from few millimeters to 40 cm (*Stapelia gigantea*). However it is only the corolla that is so diverse in size, while the reproductive parts are always relatively small, owing to functional constraints.

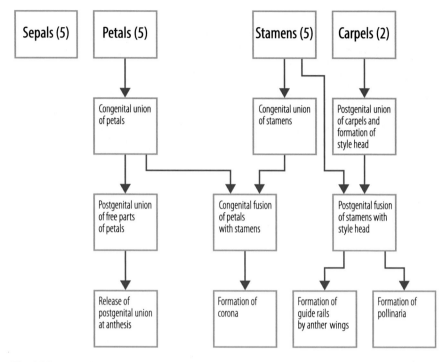

Fig. 11.9 Formation of Asclepiad flower

In order to coordinate such complex developmental processes, stability is a key prerequisite and within the Stapeliads (as well as the other ±250 genera and 3400 species of the Asclepiadaceae) the mutual position of the floral parts is very precisely fixed [253]. Hence, Stapeliads provide for an excellent model in the *physical* sense (simple, stable and reproducible) and in the *biological* sense (stable and reproducible, yet highly complex). Fusion in the corolla of Stapeliads can be described by a *geometrical* model with Gielis transformations, based on two opposing forces. The occurrence of a stable center is a direct consequence of fusion and synorganization of the petals.

Corolla Shapes in Asclepiad Flowers

These transformations impose a global anisotropy and preferred directions on $f(\vartheta)$. In other words, they 'constrain' the growth of a function $f(\vartheta)$ that wants to 'expand' or 'develop'. We call these two opposing functions the Constraining Function CF and the Developing Function DF, respectively.

When DF is a cosine function, rose or Grandi curves result, used to model certain flowers. When constrained by Gielis transformations GT they easily model a much wider variety of flowers. Indeed, by multiplying a rose curve DF with a

Constraining Function *CF*, a wide range of flowers can efficiently be described. The petals in such flowers are free standing, a condition known as choripetaly. Zygomorphy can be obtained when rose curves are inscribed in asymmetric Gielis curves (for example monogon or egg-like shapes for *m* = 1).

In sympetalous flowers petals are fused to at least some extent and the operation is not multiplication, but addition. Sympetalous flowers or Asclepiad corollas are obtained by weighted addition of the two functions *CF* and *DF*, defined by a weight parameter α with 0 ≤ α ≤ 1 (Fig. 11.10).

$$R(\vartheta) = (1 - \alpha)\left|\cos\frac{m\vartheta}{2}\right|^{n_4} + \alpha\left[\left|\cos\left(\frac{m\vartheta}{4}\right)\right|^{n_2} \pm \left|\sin\left(\frac{m\vartheta}{4}\right)\right|^{n_3}\right]^{-\frac{1}{n_1}}$$

$$= (1 - \alpha).DF + \alpha.CF \tag{10.1}$$

The influence of *DF* and *CF* (Fig. 11.11) depends on a weight parameter α. When α is close to one, petals will allmost completely be fused, like in *Caralluma*

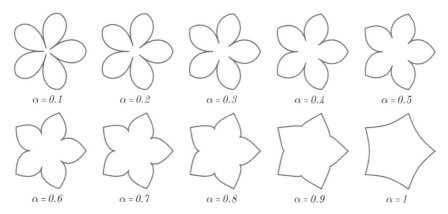

| α = 0.1 | α = 0.2 | α = 0.3 | α = 0.4 | α = 0.5 |

| α = 0.6 | α = 0.7 | α = 0.8 | α = 0.9 | α = 1 |

Fig. 11.10 Increasing fusion of petals

Fig. 11.11 **a**. *Caralluma frerei*, **b**. Fusion of *DF* and *CF*, **c**. *Huernia recondita*, **d**. constraining superpolygon

frerei. In *Huernia recondita* both functions contribute half (i.e. $\alpha = 0.5$), the arithmetic mean *AM* of *CF* and *DF* (Fig. 11.11).

A Three-Element System for Fusion in Corolla

Using the two basic operations from the *MATHS*-Box between *DF* and *CF*, the shape of any flower can be described by three strategies (*X*, *Y* and *Z*). The first strategy, *X*, defines *DF* (number and size of petals) and *CF* (constraining polygon). *Y* and *Z* involve multiplication and addition of *CF* and *DF* respectively, allowing for the description of choripetalous and sympetalous flowers. *X*, *Y* and *Z* are then combined into a three-component system, whereby *X* is present in all combinations and choripetaly and sympetaly depend on the use of addition or multiplication. *Thunbergia alata* is one example of a combination of *X*, *Y* and *Z* strategies (Fig. 11.12).

The *(X)* strategy controls the individual functions (global anisotropy *CF*, and petal size and number *DF*) and fulfills the requirements for the ultimate outcome of the *Y* and *Z* strategies, which are multiplication and addition of these functions respectively. *X* differs from *Y* and *Z*, as *A* differs from *B* and *C*. As a three-element system, *X*, *Y* and *Z* can be applied like the well-known *(A)BC* model for molecular aspects of flower development. The *ABC* model was initially defined based on the analysis of morphological mutants in only two species, treating *A, B* and *C* as independent groups of genes. MADS-box genes are a group of transcription factors, in the case of floral organs acting as homeotic genes, ensuring the development of a specific pathway towards specific organs.

Though applicable in its basics, nature should be expected to have been (and remain) flexible with regard to the genetic set up of developmental processes. Indeed, the rather diverged control of the *A* function has resulted in a revision of the *(A)BC* model. The *A*-function establishes a floral context by its involvement in the control of floral meristem identity; it provides the requirements necessary to enable the *B*- and *C*-function genes to exert their control over floral organ identity, in the very same way as *X* provides the necessary conditions for flowers to form. Under the conditions that one element always is switched on, a three element system yields only four instead of eight developmental variants, which is one of the reasons why the original *ABC* model has to be substituted by the *(A)BC model*.

Choripetaly and sympetaly in flowers can be based on these two most elementary mathematical operations from the MATHS-Box, multiplication and addition, between a fundamental function and a constraining space. This then leads to the *(X)YZ* system, which is a geometrical system, not only combinatorial. We only need the two most elementary mathematical operations from the MATHS-box and the simple relations between arithmetic *AM* and geometric *GM* means between *CF*

Fig. 11.12 Black-eyed Susan (*Thunbergia alata*)

Table 11.1 Various means in numbers, surfaces and flowers

	Geometric mean GM	Arithmetic mean AM	Weighted arithmetic mean WAM	Relations
Numbers	$\sqrt{ab}$	$(a+b)/2$	$w_1 . a + w_2 . b$	$GM \le AM$
Surfaces	$\sqrt{\kappa_1 . \kappa_2} = \sqrt{K}$	$(\kappa_1 + \kappa_2)/2 = H$	$\kappa_1 \cos^2 \vartheta + \kappa_2 \sin^2 \vartheta$	$K \le H^2$
Flowers	$\sqrt{DF.CF}$	$\frac{DF+CF}{2}$	$\alpha DF + (1-\alpha)CF$	$DF.CF \le \left[\frac{DF+CF}{2}\right]^2$

and *DF* to study flower development and evolution. These classical means are pivotal in number theory and the theory of surfaces. This is summarized in Table 11.1.

A Highly Stable Corolla

The formation of the very intricate and complex flowers requires a very stable environment. This is provided by the corolla as a stable disc, which creates a stress-free zone in the center of the corolla, and the stresses are all diverted towards the tip of the petals. In Chap. 8 we already studied the vibrations of discs and used the fusion of petals as one example. In choripetalous flowers petal responses to stress are restricted to each individual petal. In sympetalous flowers the response will be distributed in the corolla towards the periphery and in the center a stable zone is created. This central zone is essentially stress free and more fusion or synorganisation increases the size of the stable zone (Fig. 11.13).

Vibrations should be thought of as ways for the plate to get to rest as efficiently as possible. The first mode of vibrations of a *floral drum* indicates how flowers can develop with a central stable core. Singling out the first mode of vibration does not come as a surprise: the architecture of plants allows for separation of vibration

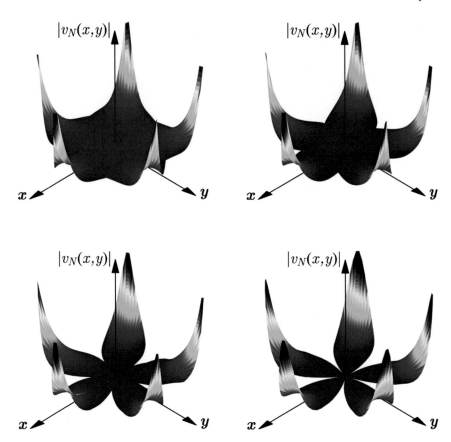

Fig. 11.13 Spatial distribution of the partial sum $u_N(x,y)$ of order N approximating the solution of the interior Dirichlet problem for the Helmholtz equation in the flower-shaped domain with fusion parameter α. Copyright Diego Caratelli

modes under wind stress and the original idea of Turing's morphogenetic patterns is to separate specific spatial and temporal modes in reaction-diffusion systems.

This biogeometrical strategy has predictive value: in this study the flower was fixed in the center with free boundaries, but when fixed at the edges, which is physically realized when the perimeter of the corolla is stiffer than the central part, the first mode of vibration resembles the 3D deformation of the corolla as observed in a variety of bell-shaped and tubular flowers. The development of cups (*Asclepiads*) and floral tubes (e.g. *Petunia*) then becomes simply the most logical way for evolution to occur from a mathematical and physical point of view. Both the corolla as a stable disc and deformations of the corolla into floral tubes are then well-defined responses to local and global stresses, both in a long time frame (*evo*) or in a shorter one (*devo*).

Fast Forward Evolution: Combining Stochastic and Deterministic Techniques

A major question is how key innovations as fusion can be stabilized and lead to specialized groups as in Asclepiads and Rafflesiads. The key innovation of large flowers with fused petals of Rafflesiaceae evolved very fast within the Euphorbiaceae with small flowers. The whole group of Asclepiads also stabilized this key innovation, and evolved into a very species-rich group. One way to understand this fast evolution is to compare developments in computer science, in particular the combination of stochastic (mutations in genetic algorithms) and deterministic (e.g. Levenberg-Marquardt) methods in image analysis [204, 205].

In the first step, the stochastic method provides one family of solutions for the raw problem of determining the appropriate symmetry and the shape coefficients without any prior information. The fine-tuning of an optimal solution is obtained in the second step using the Levenberg-Marquardt algorithm. Consequently, this strategy allows for robust and efficient shape initialization with quick and accurate determination of a solution.

This hybrid approach may help in unravelling natural evolutionary processes. Such strategies may result in very efficient strategies with short evolutionary times, combining efficient mutations with deterministic optimization strategies. Examples are the mobilization and neo-functionalization of enzymes from the large pool of enzymes, large numbers of offspring or epigenetic imprinting. Once a key innovation is formed in evolution as a broad solution, rapid diversification will follow, with adaptations not only to physical conditions, but also towards biotic factors, not in the least pollinators.

In this respect orchids (monocots) and asclepiads (dicots) share the formation of a central column, pollinaria and complex olfactory patterns. Both groups are very rich in species, resulting from the evolutionary success of the aggregation of pollen into pollinaria and the formation of column (Orchids) and gynostegium (Asclepiads). The synorganisation leading to column in orchids or the gynostegium in asclepiads is a key innovation that led to two groups very rich in species.

The main difference is the absence of fusion of tepals in orchids, but similar biophysical principles may be the basis of deformations of the labellum in orchids, when we study higher order components of the solutions. When a floral disc is mounted on a stalk, the plate vibration example showed a deflection of stress towards the pointed petal ends, creating a more stable zone. If however the condition would be that the perimeter is fixed and the whole inner domain is allowed to vibrate, then a most natural solution to deal with stress is the formation of a bell shape or tube. Many tendencies in plant development and evolution are simple solutions to relatively simple problems of optimization. Once the direction of a solution is favored it can become evolutionary stable, combining stochastic and deterministic methods.

Beyond Asclepiads

The Stapeliad corolla as a physical (stability and reproducibility) and biological system (complexity) is an excellent biogeometrical model for the study of symmetry, synorganization and stability in development and evolution in flowers. Synorganization is found in reproductive structures in over 80% of all angiosperms and is very common in vegetative leaves and stems, irrespective whether synorganised parts arise by fusion or apoptosis. Our model can be used to understand fusion as the basis of the stunning stability and floral diversity in Asclepiadaceae and the rapid evolution of giant flowers of *Rafflesia, Rhizanthes* and *Sapria* within the Euphorbiaceae; it may shed some light on the disproportionate number of species bearing gigantic flowers in sapromyiophilous species. Our model also captures the evolutionary development of oligomery and fusion, two major trends in the evolution of plants. It exhibits a very general applicability enabling to model any flower shape.

With major focus on the *MADS BOX* gene family in flower formation, progress in understanding fusion and synorganisation has been very slow. To study fusion or lack thereof in flowers *Petunia maewest* is still the major reference with reduced lateral outgrowth of initially separate primordial, failing to fuse during development. This process involves *HOMEOBOX* genes. *MAW* (*Petunia maewest*) encodes a member or the *WOX1* subfamily and the role of *MAW/WOX1* homologues in other plants seem to have similar phenotypes [243, 245, 250–252, 257, 258]. This does shed some light on fusion, but a thorough study and understanding will need a multifaceted approach, including physiology, biophysics and geometry [258] on various plant models beyond *Arabidopsis*, such as *Petunia* [257] and *Medicago* [259].

With our biogeometrical approach a framework for understanding fusion in plants is opened, helping to understand the ubiquity of synorganization in plants by fusion between different parts and organs, as a major driving force in the evolution of angiosperms [248, 249]. In fact spiral versus whorled growth is also a major force in how plants are related phylogenetically. All lower plants, up to early diverging monocots and dicots (such as Magnolia and water lily) show spiral growth and spiral reproductive structures. Eudicots on the other hand have in general, whorled phyllotaxy. The Lamiids are characterised by sympetaly, and two main subgroups are distinguished by early or late sympetaly [248, 260]. In rosids flowers are choripetalous, but nevertheless flowers succeed in building a diversity of floral tubes due the ingenious way in which the lower parts of the petals have special ribs and overlaps, as was shown in *Geranium* [261]. One thing is certain: nature is very inventive and plants, from unicellular algae to giant trees, have explored a wide variety of forms successfully colonizing a wide variety of habitats, from a relatively compact genetic toolbox. Evolving from small euphorbia flowers to giant Rafflesia flowers, or changing symmetry of perianths, many solutions have been developed and are reused if needed. Perianth symmetry has changed at least 199 times during angiosperm evolution [262]. The Orchids (monocots) and Apocynaceae (dicots) have evolved similar solutions, which led to species rich

Fig. 11.14 Flowers are key to understanding nature. Copyright Fabian Gielis

groups in both cases [240]. Recent advances in next "supermodels" like *Petunia* [263] will lead to rapid advances in understanding.

We also emphasize that synorganization is a broad phenomenon. For example, the cardioid analogy applied to choripetalous and sympetalous flowers then allows for modeling a wide range of composite or palmate leaves with differing degrees of fusion, as can be observed within the genus *Acer*. In an evo-devo perspective, understanding flower formation and evolution as the sound of drums favors the idea of a co-active role of plants and pollinators in co-evolution. The weighted arithmetic mean involves a close cooperation between two different entities. These may be fusion of two organs, but also 'fusion' between outside and inside. Here outside may be environment and inside the natural shape in developmental or evolutionary sense.

The combination of trigonometric functions with Gielis transformations, with the exact solution to boundary value problems in arbitrary domains, as applied here with the Helmholtz equation for floral development and morphology, can in fact be used in any field of science. Flowers are just one graphical expression of the equations used (Fig. 11.14). If we express the equations in time diagrams, our flowers will turn into waves.

Part VI
Συμπέρασμα—Conclusio

Chapter 12
The Pythagorean Theorem for the Third Millenium

Retreat to Euclid.

A.N Alexandrov

Preserving the Pythagorean Theorem...

One uniform, Pythagorean-compact description with all the associated character-istics that come for free allows for the study of natural shapes in a natural way. We have strayed not too far from common ground using the normal rules of arithmetic and circular (trigonometric) functions, and the total is compacted into a Pythagorean structure. The genius of Pythagoras was, I repeat, the transformation of circular motion into rectangular motion based on areas, and doing so by avoiding geometric means.

The Gielis Formula describes a wide range of abstract and natural forms and shapes, in the compact Pythagorean structure:

$$\varrho(\vartheta) = \frac{1}{\sqrt{\left[|\cos(\vartheta)|^2 + |\sin(\vartheta)|^2\right]}}$$

We readily find the generalized trigonometric functions $_p\cos\vartheta$ and $_p\sin\vartheta$ associated with the shape and the corresponding generalized Pythagorean Theorem (for Lamé curves; this can easily be extended to more general functions as well) and conservation laws (area in the case of rectangular triangle). It is a simple and straightforward generalization.

Let us look at this Pythagorean structure in a different way. In the transition from Pythagorean Theorem $a^2 + b^2 = c^2$ to the general equation of a circle, specific numbers are substituted by variables x, y to arrive at $x^2 + y^2 = R^2$. The variables x, y are "placeholders" which can be substituted by certain numbers. A $XY-$graph of the circle results when for all variables x we find the corresponding variable y for which the given equation holds.

© Atlantis Press and the author(s) 2017
J. Gielis, *The Geometrical Beauty of Plants*,
DOI 10.2991/978-94-6239-151-2_12

Placeholder: (a) input fields that are empty or with provisional text or image, until the field is filled with the relevant information for final use. (b) A symbol or piece of text used in a mathematical expression or in an instruction in a computer program to denote a missing quantity or operator.

In 1818 Gabriel Lamé generalized the circle: instead of 2, he made the exponent into a variable n, which gives Lamé curves, more specifically supercircles. Squares, circles and intermediate shapes differ only in the variable n:

$$|x|^n + |y|^n = R^n,$$

Using "placeholders" instead of numbers or variables, yields:

$$|x|^{\blacksquare} + |y|^{\blacksquare} = R^{\blacksquare},$$

When we select the value $\blacksquare = 2$ we have the circle, if we select $\blacksquare = 1$ the inscribed square results. For $\blacksquare = 3$ or $\blacksquare = 5$ we get supercircles, inflated compared to the circle. We can choose any value (any real number) between 0 and infinity, and thus morph a square seamlessly into a circle or supercircle.

If the radius of the circle is 1, we can use, (instead of x, y) the cosine and the sine of the angle ϑ. The radius is always equal to one and the Pythagorean Theorem gives $\cos^2 \vartheta + \sin^2 \vartheta = 1$.

Converting Lamé's equation into polar coordinates we arrive at:

$$\varrho(\vartheta) = \frac{1}{\sqrt[n]{|\cos(\vartheta)|^n + |\sin(\vartheta)|^n}}$$

Again using "placeholders" instead of numbers or variables, in this case two different placeholders, $\blacksquare$ for n and $\ldots$ for ϑ we have:

$$\varrho(\vartheta) = \frac{1}{\sqrt[\blacksquare]{|\cos(\ldots)|^{\blacksquare} + |\sin(\ldots)|^{\blacksquare}}}$$

For $\blacksquare$ we can fill in n, or three different numbers n_1 or n_2 or n_3. We can also choose to introduce some function $f(n_1), f(n_2), f(n_3)$ whereby the value of n_1, n_2 and/or n_3 changes along the shape or over time, for example evolving from circle to a starfish.

For the placeholder $\ldots$ we can fill in ϑ (Lamé curves, and the circle if $n = 2$), or $\frac{m}{4}\vartheta$ (Gielis curves and transformations), or $f(\vartheta)$ for the generalization of Gielis curves we introduced earlier. Within these two, $\blacksquare$ is the shape factor and $\ldots$ is the symmetry factor. The former can morph a circle into starfish or a square and back, the latter defines the symmetry or spacing of the angle, and whether we get a square or a starfish.

The result is that $\varrho(\vartheta)$ becomes a variable of stretchable radius, whose length is dependent on the actual numbers used in the placeholder $\blacksquare$, and the spacing of the radii depends on $\ldots$. With those two we can morph a circle into any of the shapes shown in this book, and a zillion more.

…With Elementary Mathematics

We have used only elementary mathematics. Foundations laid by ancient Greeks, basic rules of arithmetic, games of cubes and beams, conic sections and a modern version of the Pythagorean Theorem are all the ingredients needed to arrive at a mathematical technique, expanding and generalizing Lamé's calculus, useful for studies in geometry and the natural sciences (and also arithmetic, algebra and so on).

One should not confound elementary and basic, with superficial. On the contrary: a deeper understanding of the foundations of the inheritance from Ancient Greek mathematics led us to these developments. Greek mathematicians themselves built on earlier findings from other cultures, but they not only added the idea of deduction and proofs, but also strived for a description as simple as possible. This became the circle and conic sections, projectively equivalent to the circle.

These conic sections could be derived from the application of areas and the various means (AM, GM, HM) of the Pythagoreans, and were studied in detail by one of the greatest mathematicians of ancient history, Apollonius of Perga. Twenty centuries after Pythagoras, these conic sections formed the basis of our scientific revolution(s), through Kepler, Galilei, Newton and Fourier, to name a few.

The tendency has largely been to neglect the source, and focus on expansions and novelties. Instead of the source and the river that sprang from it, we now see only the delta, an intricate system of small rivers, creeks and canals. We find ourselves in a Tower of Babel situation with each subfield continuously fighting or struggling to prove and improve its importance. In each direction we find hyper-specialists, with high-level languages that are only spoken and understood by a few and among all fields it seems that algebra and algorithm have taken the lead, neglecting the geometrical foundations. What can easily be understood by geometrical reasoning is reduced to algorithms, algebra, all types of new kinds of sciences, erasing the past.

The Geometree however, is very much alive and thriving, as we showed. The foundations the ancient Greek provided are deep and fundamental, the river that sprang from this source continues to provide millions of tons of fresh water to our delta. In this sense, all mathematics and its applications in the natural sciences should be elementary, built on solid Elements or Στοχεῖα.

Preserving Euclidean Geometry

Considering the more general transformations instead of the various instances for suitable choices of parameters, we add semi-rigidity to Euclidean geometric transformations (translation, rotation, reflection, glide reflection and scaling). Defining unit circles in a different way in Euclid's set of definitions and thinking of circle, square and starfish as commensurable figures, Euclidean geometry can be generalized greatly, encompassing many earlier results in mathematics.

Natural organisms, shapes and forms are Euclidean objects then and widely varying simple and complex objects, forms and shapes, forming the complex set of abstract or natural objects, are now equally simple. It corroborates Thom's dictum that *"The dilemma posed by all scientific explanation is this: magic or geometry. Classical Euclidean geometry can be considered as magic; at the price of minimal distortion of appearances (a point without size, a line without width) the purely formal language of geometry describes adequately the reality of space. We might say, in this sense, that geometry is successful magic"*.

A continuous transformation connects different unit circles that are found in nature and we remain within the Euclidean framework. We expanded the definitions of curvature, based on the definition itself, in the spirit of classical Euclidean geometry where the circle is the measure of curvature.

With our description of shape we can model a wide variety of natural shapes. This description of shape can be a most compact description, in the Pythagorean-compact sense, a best fit description in Chebyshev' sense, yielding optimal coordinate systems or garment adapted to the shape in the sense of Lamé, or best model in the Aikake sense. It can be used to model various natural processes, comparable to or better than current methods. Its capacity to transform shapes widens the capabilities of current methods, many of those dating back to the 18th and 19th century.

While providing various new methods and tools, we are essentially describing and modeling nature. Even if the models are better, we understand the limitations of man in a way similar to Euler. Vladimir Arnold wrote: *"The mathematical technique of modeling consists of ignoring this trouble and speaking about your deductive model in such a way as if it coincided with reality. The fact that this path, which is obviously incorrect from the point of view of natural science, often leads to useful results in physics is called "the inconceivable effectiveness of mathematics in natural sciences" (or "the Wigner principle")"* [8].

What if?

In 2018 we celebrate the bicentennial of Lamé curves; they are known as superellipses and supercircles, but deep down they are the geometrical equivalent of the Last Theorem of Fermat and a generalization of the Pythagorean Theorem with a direct link to the study of nature. They originate in the study of natural shapes since Lamé found these curves excellent for applying geometrical methods to the study of crystals. Gabriel Lamé clearly understood the generality of his findings, but also recognized their limitations. Gielis curves are only two decades old, but they apply the ideas of Lamé for any symmetry, also motivated by the desire to describe and study natural shapes. In this way they make Lamé's new calculus applicable to a much wider range of abstract and natural shapes. We find that Gielis curves provide excellent models for the study of natural shapes and phenomena.

The only strange aspect is that they are separated by 180 years, whereas all ingredients were available in the second decade of the 19th century, with Lamé and Gauss. When this was realized at that époque, Fourier's results could have been generalized to any domain already then, instead of 200 years later. Gauss would have noticed the correspondence with his elliptic functions and arithmetic-geometric mean. Time dilatation and increasing weight with increasing speed in Special Relativity Theory would not have been strange at all, but only one demonstration of the "Lamé" fingerprint in nature. The study of natural shapes would have been possible with their specific trigonometric identities and Pythagorean Theorem.

Riemann would have given the fourth power as distance function more consideration and would have recognized its significance from a geometrical point of view. What Riemann felt correctly, at least in one sense of the word, is that this would not contribute much new. Indeed, by our transformations everything can be mapped back to the circle and trigonometric functions, corroborating the excellent work of older mathematicians. He was not correct in at least one other sense: it provides us with new glasses to study nature, its shapes and its phenomena.

Deterministic and Structurally Stable

There is also a "beyond" view. I believe it allows us to look at mathematics and its role in modeling natural processes quite differently, changing the way we look at nature, and at ourselves as human beings, in relation to the world we live in, sharing ecosystems with other beings of organic or inorganic nature. How then could this influence our current worldviews?

When Pythagoras envisioned and taught his doctrine '*All is number*', the circle became the cypher of the world, together with the conic sections. From the very moment that Gabriel Lamé conceived his book of 1818, all Lamé curves, with their direct connection to many fields in mathematics, were completely defined in their 'final' form. The generalization of Lamé curves into Gielis curves and surfaces is completed since 1997. The connection with many other fields, outlined in this book, will provide food for thought, in particular transformations of Euclidean forms and semi-rigid geometry. Its major "beyond" impact could well be in bridging the differential and the difference—the continuous and the discrete—using more general geometric definitions of differentiation and manifolds, combining the smooth and piecewise linear.

Another way how it could influence our views is that we can think more deterministically, less governed by pure chance. It is clear that our world is not chaotic or random, but a cosmos, well arranged and ordered. Certainly it is dynamic and moving all the time, but a certain order and harmony is the imprint of nature. Even when our dynamical systems show fingerprints of chaos, one can use Lamé-Gielis curves to increase structural stability. One example: the transition to chaos with the logistic equation may be damped by slightly modifying the exponent of the logistic equation from 2 to 2.3 or so; chaos does not set in so easily. Generalized Möbius-Listing bodies, cut a certain number of times, facilitate the study of various dynamic systems, the structure and the beating of a heart, the formation of galaxies and planetary systems, or the behavior of elementary particles.

Our cosmos is by and large deterministic, rather than based on chance and contingency. Hence, our worldviews can also become more deterministic. Chance has led to remarkable progress in science especially wherever chance is used in a way to describe microscopic systems governed by chance, leading to structurally stable systems at the macroscopic level. Our world is structurally stable.

Let us be clear: A worldview based on pure chance is a priori anti-scientific. Acknowledging that there are a number of natural phenomena that we can never model is giving up on understanding (resorting to computations at most). This belief in pure chance is entirely negative, without any scientific interest. René Thom wrote: «*Le hasard est un concept entièrement négatif, vide, donc sans intérêt scientifique. Le déterminisme en science n'est pas une donnée, c'est une conquête. En cela, les zélateurs du hasard sont les apôtres de la désertion*» [264]. Determinism in science is not a given but the real victory of science. Also laws of probability or Kolmogorov-Sinai entropies are deterministic.

Geometry is fundamental in any scientific description of the world, and with our continuous transformation (or subsets) we can describe many more phenomena and shapes without having to resort to pure chance or pure randomness as the ultimate foundations of natural phenomena.

Plants at the Core of Our Worldview

The idea of Gielis transformations originated in the study of plants, in particular of square bamboos and the hollow culms of bamboo. This led to a Pythagorean compact description of a variety of natural and abstract shapes, describing the symmetry and dimension of simple and complex shapes in plants. From simple algae and diatoms, to the lower and higher plants, two to six parameters suffice for the accurate description and quantification of the shape of plant organs, tissues and cells. For the modular architecture of plants, a logico-geometrical methodology was presented, based on R-functions, whereby domains could be described efficiently and studied quantitatively. Such domains could be layers of onions or bamboo shoots, or annual rings in trees. Modular architecture is also the efficient stacking of vessels in wood, or the packing of cells in sporangia.

The two basic shapes in nature, circle and spiral, can be discerned everywhere, in the growth of organs, the transition of vegetative to generative organs, the phyllotaxy in plants, up to the evolutionary pathways, from spiral polymery to whorled oligomery. A geometrical model was proposed to study fusion of petals in Asclepiads, whereby the formation of a floral tube, either resulting from true fusion or from clever mechanisms in Geranium, is a very natural consequence of vibrations affecting the flower. Several methods were proposed to study plants in the

Table 12.1 Allometry simplified

Variables x, y, power n, m	Sum	Product
Result	$|x|^n + |y|^m = R^p$	$|x|^n = |y|^m$
Means	Arithmetic means	Geometric means
Curves	Lamé curves	Superparabolas
Effect	Shape and size	Power laws

very same way as various physical phenomena are studied. The highly improved accuracy to model plant parts results in more accurate allometric laws, based on geometric rather than arithmetic means. This is summarized in Table 12.1, and can easily be extended using Gielis transformations for any study in the natural sciences.

Physis-Centered Worldviews

In understanding natural objects and phenomena our scientific methods and worldviews are by and large based on our human modeling, with circular glasses to look at nature; these methods are Euclidean or Euclidean in disguise, having straight lines and circles as the main focal points in whatever we do. However, these glasses are our bulls in the collections of delicate porcelain and filigree, which nature is. We cast, very "anthropomorphically" and "anthropometrically" our human nets over nature, but fail to understand it. We should start thinking in geometries adapted to the shape. With our approach, we provide for unit circles adapted to the natural shapes. If starfish would develop geometries based on their experience of this world, their unit circle may well be the starfish itself. And the shape of petals in flowers is related to the curvature imposed by the unit circle of early developing flower buds.

We now have new glasses, microscopes and telescopes that allow us to look at nature beyond our myopic anthropomorphic instruments and the good news is that such unit circles are members of a continuum of unit circles. They can all be connected to our classic circles and circular functions and we can extend our Euclidean geometry with our semi-rigid transformations. We can develop natural coordinate systems onto natural objects and phenomena, following the vision of Gabriel Lamé and his dream of a Unique Rational Science.

Human-Centered Worldviews

We have argued earlier how Lamé curves focus on pure numbers avoiding the use of geometric means. In the evolution of science geometric means have led to the classical "statistical" approach, with laws of large numbers etc., and we classify objects in classes or topoi, and define relations between them. Persons X, Y and Z may be in the same class, based on 98% of commonalities of say four traits. Unfortunately, these methods are not used to understand humans, but predominantly to exert control over groups and populations.

Mathematics has thus become the "science of patterns", of commonalities. This is incomplete: mathematics should and could be, like natural language (including art and poetry), a way of expressing very deep feelings and individual aspects. Persons X, Y and Z may have three or five aspects in common, but they are different in a million more. We should not even call them X, Y or Z. All leaves on a tree are different, and so are all cells inside an organism. Together with all other elementary particles, electrons are mathematical abstractions, but for any sane person it is impossible to believe that every electron, or every proton in the whole universe is exactly the same.

The basis of Lamé curves and Gielis transformations is avoiding geometric means and I have argued that this new view of pure numbers is not antagonistic to the current world-views, but dual. It is a way of describing accurately a very specific sequence or series, without having to resort to stochastic systems. We are now able to assign individual parameters to individual starfish, or to parallel cross sections of cacti, or to developing shapes from circle to pentagon. Why not develop a mathematical language to address the individuality in multiplicity, manifolds and *Mannigfaltigkeiten* that can describe individual people, or any individual member of the large catalogue of Universal Natural Shapes? The next natural step is then to find the ratios that harmonize the co-existence of individuals.

The Sixfold Way

We followed the rigorous scheme of building as developed by the ancient Greek.

πρότασις *or Propositio* exposed what needs to be shown, namely that beyond the mere analogy, we can develop a rigorous geometrical and mathematical approach to study natural shapes based on Lamé-Gielis curves and surfaces.

Εκθεσις *or Expostio*: From elementary notions in mathematics that are at work in every field of mathematics and its applications in the natural sciences, we arrived at Lamé-Gielis curves and surfaces.

Διορισμός: in this determinatio step we showed how shapes and their combinations can be firmly embedded into mathematics. Gielis is an acronym for Generalized (or Geometric) Intrinsic and Extrinsic Lengths in Submanifolds. Length refers to anything that can be measured and applies to a wide variety of natural shapes and phenomena, for the study of everything we wish to understand— πάντα γα μὰν τὰ γιγνωσκόμενα.

Κατασκευή *or Constructio*: Various methods have been presented, from the Generalized Gielis formula, over R-functions to analytical solutions of boundary value problems, using classical 19th century methods. Defining curvatures, classical via circles in the Oresme-Huyghens-Newton tradition, or based on Lamé curves and supershapes, we construct powerful tools to study natural shapes. The relation between Rhodonea curves and Chebyshev polynomials allowed for rewriting the Gielis Formula.

Απόδειξις or *Demonstratio*: Lamé and Gielis curves are excellent models for studying various natural shapes, e.g. in bamboo leaves and tree rings. They provide for an elegant and powerful way of representation for far-from-equilibrium shapes, as in snowflakes and DNA. These models and their relation to old notions in geometry help understanding evolutionary and developmental aspects in biology, and fusion in Asclepiad corollas was used as example.

Συμπέρασμα or *Conclusio*: Many of the ideas from the original book Inventing the Circle have been refined and turned out to be very useful. *Description always precedes understanding* the connections between mathematics and nature. Moreover, Lamé-Gielis curves and Gielis transformations open the door for many new developments.

Among the various ideas and methods presented, I would like to mention three main results: First, an "existence proof" that natural shapes can be described in a Pythagorean-compact and Pythagorean-simple (oligomial) way, generalizing the circle. Second, Lamé's calculus, which Lamé himself considered to be difficult, is now made general and easy, and is fully embedded in the history and the main body of mathematics and contemporary geometry, with *Generalized* (or Geometric) *Intrinsic and Extrinsic Lengths in Submanifolds*, acronym Gielis. Third, the expression of the Gielis Formula with Chebyshev polynomials links geometry and algebra.

An Open Invitation to Mathematicians and Students of Nature

When the Gielis Formula was published initially as Superformula, generalizing Lamé-Hein' superellipses, and Barr's Superquadrics, this was met with enthusiasm from within mathematics, and it was renamed as Gielis Formula, by mathematicians. This formula can be considered as Gielis Transformations, giving rise to Gielis curves, surfaces and submanifolds. In the past decade we have come to understand that it fits well into the long tradition of mathematics, broadening the applications of the foundations laid down by the Ancient Greek mathematicians. As is shown in the book, it is in fact, through Lamé equations, a generalization of circle and the Pythagorean Theorem. We may consider this as a modern version of the Pythagorean Theorem, broadening the scope to a wide range of natural and abstract shapes, with a new pair of glasses. A Pythagorean Theorem for the Third Millenium.

Great challenges lie ahead. The geometrical foundations underlying all natural laws of nature need to be combined with many of the observations from chemistry, molecular biology, physiology and biochemistry, as well as from the field of mechanics (dynamics and kinematics). We need to build bridges with *"seasoned experimentalists used to working with living matter and always struggling with an elusive reality"*, to finally arrive at *"an abstract, purely geometrical theory of morphogenesis, independent of the substrate of forms and the nature of the forces that create them"*. This geometrical theory of morphogenesis applies to all natural shapes at all levels, not only living matter.

We need not to stray too far from Euclidean geometry, since this is what we know and experience as human beings. I wish to add three crucial aspects. First of all, observations are crucial, and one should at all times be aware that our models are based on a theoretical framework that is always somewhat biased, even if 99.99% of scientists agree. For the observation of phenomena we should not limit ourselves to our inner scientist circles. Goethe's observations turned out to be crucial for understanding plants from a dynamical perspective [265]. There are shapes inbetween circle and square but few realize that Lamé curves exist (remember Fig. 2.3).

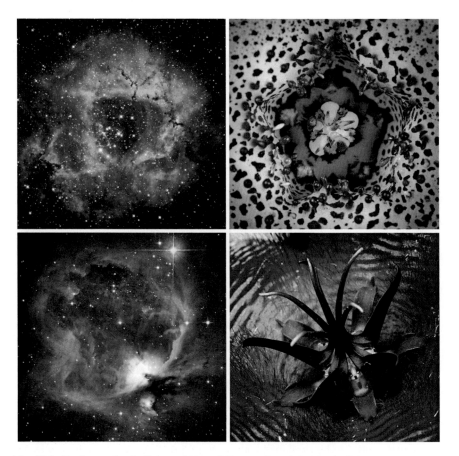

Fig. 12.1 Fatal attraction in Universal Natural Shapes. Copyright Martin Heigan

Second, every human being is different, with his or her own talents and time needed to process and understand information. Third, the rules of rational thinking are incomplete. Our observations led us to different glasses, micro- and telescopes to look at nature. The complex turned out to be simple. Human imagination is what distinguishes us from computers, and what we should honour in education.

To B.L. van der Waerden's *"Every branch of mathematics can exist by itself as a logical system. When we wish to develop mathematics as a living and growing science, we can understand this only in the symbiosis with physics, and astronomy. Only in this symbiosis can our beloved science blossom, grow and forever stay young"* [266], we must add *"biology and botany"*. Fortunately the close relationship between *Theory and Practice* (Simon Stevin's *Spiegeling en Daet*), has always been pivotal in all of scientific developments and breakthroughs. *Geometry, and botany and nature in general*, are *successful magic* (Fig. 12.1).

References

1. Gielis, J. (2001). *De uitvinding van de Cirkel*. Antwerpen, België: Geniaal. ISBN 90-6215-792-0. Gielis, J. (2003). *Inventing the Circle*. Belgium, Antwerp: Geniaal Publishers.
2. Gielis, J. (2003). A generic geometric transformation that unifies a wide range of natural and abstract shapes. *American Journal of Botany*, 90(3), 333–338.
3. Verstraelen, L. (2004). Letter sent to journalists.
4. Philips, T. (2003). The Superformula. Math in the Media. Retrieved from http://www.ams.org/news/math-in-the-media/mmarc-05-2003-media
5. Heath, T. (1981). *A history of Greek mathematics* (Rev. Ed.). New York: Dover Publications.
6. Berger, M. (2003). *A panoramic view of Riemannian geometry*. Springer Verlag. doi:10.1007/978-3-642-18245-7
7. Arnold, V. I. (1998). On teaching mathematics. Extended text of the address for the discussion on teaching of mathematics in Palais de Découverte in Paris on 7 March 1997.
8. Lamé, G. (1818). *Examen de différentes méthodes employées pour résoudre les problèmes de géometrie*. M. V. Courcier imprimeur Libraire, Paris.
9. Thom, R. (1972). *Structural stability and morphogenesis*. Benjamin.
10. May, R. M. (2004). Uses and Abuses of Mathematics in Biology. *Science*, 303, 790–793.
11. Schrödinger, E. (1940). The general theory of relativity and wave mechanics. *Wis- en natuurkundig Tijdschrift*, 10(1), 2–9.
12. Verstraelen, L. (2005). Universal Natural Shapes. *Journal for Mathematics and Informatics of the Serbian Mathematical Society*, 40, 13–20.
13. Koiso, M., Palmer, B. (2008). Equilibria for anisotropic energies and the Gielis Formula. *Forma* (Society for Science on Form, Japan), 23(1), 1–8.
14. Natalini, P., Patrizi, R., & Ricci, P. E. (2008). The Dirichlet problem for the Laplace equation in a starlike domain of a Riemann surface. *Numerical Algorithms*, 49(1–4), 299–313.
15. Matsuura, M. (2015). Gielis' superformula and regular polygons. *Journal of Geometry*, 106 (2), 1–21.
16. Chern, S. S. (2000). Introduction. As cited in: Dillen, F. & Verstraelen, L. (eds.), Handbook of differential geometry. Netherlands, Amsterdam: Vol. 1.
17. Weil, A. (1978). S.S. Chern as Geometer and Friend. As cited in: *Chern, S.S. (1978). Selected Papers*. Springer Verlag.
18. Serrano, I. M. & Suceava, B. D. (2015). A Medieval Mystery: Nicole Oresme's Concept of Curvitas. *Notices of the AMS*, 62(9).
19. Chern, S.S. (1990). What is geometry? *The American Mathematical Monthly*, 97(8), Special Geometry Issue, 579–686.
20. van der Waerden, B.L. (1983). *Geometry and algebra in Ancient civilizations*. Springer Verlag.

© Atlantis Press and the author(s) 2017
J. Gielis, *The Geometrical Beauty of Plants*,
DOI 10.2991/978-94-6239-151-2

21. Pauli, L. (1979). Le colloque de Royaumont. *Math Ecole*, 90, 3–10.
22. Kutateladze, S. S. (2005). *AD Alexandrov: Selected Works Part II: Intrinsic Geometry of Convex Surfaces*. CRC Press.
23. Kutateladze, S. S. (2012). Alexandrov of ancient Hellas. *Siberian Electronic Mathematical Reports*. Сибирские электронные математические известия, 9, A6–A11.
24. Rainich, G.Y. (1932). *Mathematics of Relativity. Lecture Notes*. Edwards Brothers.
25. Euler L. (1779). On the nature of air. http://eulerarchive.maa.org//pages/E527.html
26. D'Arcy, T. W. (1917). *On Growth and Form*. U.K., Cambridge: Cambridge University Press.
27. Courant, R. (1950). *Dirichlet's principle, conformal mapping and minimimal surfaces*. New York: Interscience.
28. Thompson, A. C. (1996). *Minkowski geometry*. Encyclopedia of Mathematics and its Applications Vol. 63. U.K., Cambridge: Cambridge University Press.
29. Poincaré, H. (1907). *The value of science*. The Science Press, New York.
30. Huxley, J. S. (1932). *Problems of relative growth*. New York: Dial Press.
31. McClintock, B. (1950). The origin and behavior of mutable loci in maize. *Proceedings of the National Academy of Sciences*, 36, 344–55.
32. Keller, E.F. (1983). *A feeling for the organism. The life and works of Barbara McClintock*. New York: W.H. Freeman and Company.
33. Thom, R. (1991). *Prédire n'est pas expliquer*. Paris: Champs Flammarion.
34. Osserman, R. (1996). *Poetry of the Universe – a mathematical exploration of the cosmos*. New York: Anchor Books.
35. Thurston, W. P. (1998). On proof and progress in mathematics. *Bulletin of the American Mathematical Society* 30(2), 162–177.
36. Verhulst, R. (1982). Nomograms for the Calculation of Roots, *Congres Internationale pour l'Enseignement des Mathematiques. CIEM; Belgian subcommittee of ICMI*, 193–203.
37. Verhulst, R. (2006). *In de ban van de wiskunde*. België, Antwerpen: Garant.
38. Verhulst, R. (1982). Wiskunde en Onderwijs, 29, 95–123.
39. Gielis, J., Verhulst, R., Caratelli, D., Ricci, P. E., & Tavkhelidze, I. (2014). On means, polynomials and special functions. *The teaching of mathematics* 17(1), 1–20.
40. Pogorelov, A. V. (1980). *Geometry*. Russia, Moscow: Mir publishers.
41. Crone, E., Dijksterhuis, E.J., Forbes, E.J., Minnaert, M.G.J., & Pannekoek, A. (1958). *The Principal Works of Simon Stevin*. The Netherlands, Amsterdam: c.v. Swets & Zeitlinger.
42. Verstraelen, L. (2014). A concise mini history of geometry. *Kragujevac Journal of Mathematics*, 38(1), 5–21.
43. Sarton, G. (1934). Simon Stevin of Bruges (1548-1620). *Isis* 21(2), 241–303.
44. Sarton, G. (1935). The first explanation of decimal fractions and measures (1585). Together with a history of the decimal idea and a facsimile (no. XVII) of Stevin's Disme. *Isis*, 23(1), 153–244.
45. Bosmans, H.S.J. (1926). Le mathématicien belge Simon Stevin de Bruges (1548–1620). *Periodico di Matematice* serie IV, vol VI(4), 231–261.
46. Bosmans, H.S.J. (1923). Le Calcul infinitésimale chez Simon Stevin. *Mathesis*, 37, 12–19.
47. Gielis, J., Caratelli, D., Haesen, S., & Ricci, P.E. (2010). Rational mechanics and Science Rationelle Unique. As cited in: Paipetis, S., Ceccarelli, M. (Eds.) *The Genius of Archimedes: 23 centuries of influence on mathematics, science and engineering*. Springer Verlag, HMMS, 11, 29–43.
48. Bosmans, P. (1923). Sur l'interprétation géométrique donnée par Pascal à l'espace a quatre dimension. *Annales de la société scientifique de Bruxelles*, 42, 337–45.
49. Viète, F (1591). Introduction to the analytic art. English translation as appendix in: Klein, J. Greek mathematical thought and the origin of algebra. Dover Publications 1968.
50. Colson, J. (1732) Introduction. In: Newton, I. Method of Fluxions and infinite series with its application to the Geometry of Curved lines. U.K., London: Henry Woodfall.
51. Stevin, S. (1965). *De Thiende*. Dutch Classics on History of Science (1585).

52. Van der Waerden, B.L. (1985) *A history of algebra. From al-Khwārizmī to Emmy Noether.* Springer Verlag
53. Stepanov, A. A., & Rose, D. E. (2014). *From mathematics to generic programming.* Pearson Education.
54. Newton, I. (1732). *Method of Fluxions and infinite series with its application to the Geometry of Curve-lines.* U.K., London: Henry Woodfall.
55. Newton, I. (1952). Mathematical Principles of Natural Philosophy, Great Books of the Western World, Volume 34. Encyclopedia Britannica Inc & William Benton Publisher.
56. Gielis, J. (2002). Exponent moderation in plants. As cited in: Kumar, A., Rao, I.V.R., Sastry C. (2002) Bamboo for sustainable development. Utrecht, Boston, Tokyo: VSP Publishers.
57. Niklas, K. J. (1994). *Plant Allometry. The scaling of Form and Process.* Chicago: Chicago University Press
58. Huxley, J. (1932). *Problems of Relative Growth.* Baltimore: Johns Hopkin University Press.
59. Niklas, K. J. (2004). Plant allometry: is there a grand unifying theory?. *Biological reviews,* 79(04), 871–889.
60. Dattoli, G. B., Germano, M.R., Martinelli, P.E., Ricci. (2009). Monomiality and partial differential equations, *Math. Comput. Modelling,* 50(9–10), 1332–1337.
61. Dattoli, G. (1999). Hermite-Bessel and Laguerre-Bessel functions: A by-product of the monomiality principle. As cited in: Cocolicchio, D., Dattoli, G., & Srivastava, H.M. Advanced Special Functions and Applications. Proc. Melfi School on Advanced Topics in Mathematics and Physics, 147–164.
62. Berlingeri, C., Dattoli, G., Ricci, P.E. (2007). The monomiality approach to multi-index polynomials in several variables, *Georgian Math. J.,* 14(1): 53–64 (2007).
63. Dattoli, G. (2000). Generalized polynomials, operational identities and their applications. *Journal of Computational and Applied mathematics,* 118(1), 111–123.
64. Prvanovic, M. (2017). Farkas and Janos Bolyai. Topics in Modern Differential Geometry, 1, Atlantis-Springer.
65. Gardner, M. (1975). *Mathematical Carnival.* New York: Alfred A. Knopf.
66. Gielis, J. (2009). The geometry of universal natural shapes. In: Lestrel, P. E. (2011). Biological Shape Analysis: Proceedings of the 1st International Symposium. Tsukuba, Japan: World Scientific.
67. Russo, L. (2004). *The forgotten revolution. How science was born in 300BC and why it had to be reborn.* Germany, Berlin: Springer Verlag.
68. Dattoli, G. (2000). Introduction. As cited in: Cocolicchio, D., Dattoli, G., & Srivastava, H. M. (2000) Advanced Special Functions and Applications. Proc. Melfi School on Advanced Topics in Mathematics and Physics, 147–164.
69. Feynman, R.P. (1965). *The character of physical law.* Modern Library.
70. Atiyah, M. (2000). Mathematics in the 20th century. Fields Lecture June 7–9, Toronto.
71. Loria, G. (1902). *Spezielle algebraische und transscendente ebene Kurven: Theorie und Geschichte.* Teubner G.B. Verlag.
72. Texeira, F.G. (1905). *Tratado de las Curvas Especiales Notables.* Madrid: Imprenta de la "Gaceta de Madrid".
73. Euzet, M. (1835 & 1854). Sur les courbes planes à équations trinômes et les surfaces à équations tétranômes. *Nouvelles Annales de Mathématiques.*
74. Klein, F. (1932). *Elementary mathematics from an advanced standpoint: Geometry.* McMillan, New York.
75. Barrow, I. (1916). *The geometrical lectures of Isaac Barrow* (translated by J.M. Child). Phoenix Edition.
76. Comte, A. (1843). *Traité élémentaire de géometrie à deux et trois dimensions.*
77. Gauss, C.F. (1861) in: *Gauss Werke,* Vol. 2, 629.
78. Greitzer, S. (2008). Gabriel Lamé. Complete Dictionary of Scientific biography. Charles Scribner's Sons

79. Bertrand, J., (1870) Funérailles de M. Lamé. *Bulletin des sciences mathématiques et astronomiques*, 1, 189–195.
80. Barbin, E., et al. (2009). Gabriel Lamé: les pérégrinations d'un ingénieur du XIXe siècle. *Actes du Colloque. Sabix bulletin*, 44.
81. Coolidge, J.L. (2003). *A history of geometrical methods*. Dover Phoenix Editions.
82. Lucas, E. (1891). *Théorie des nombres*. France, Paris: Gauthier-Villars.
83. Kolmogorov, A.N., Yushkevic, A.P. (1996). *Mathematics of the 19th century: Volume 1, Volume 2 & Volume 3*.
84. Lamé, G. (1859). *Leçons Sur Les Coordonnées curvilignes et Leurs Diverses Applications*. France, Paris: Mallet-Bachelier.
85. Guitart, R. (2009). Les coordonnées curvilignes de Gabriel Lamé, réprésentations des situations physiques et nouveaux objects mathématiques. In: Barbin, E., et al. (2009). Gabriel Lamé: Les pérégrinations d'un ingénieur du XIXe siècle. *Actes du Colloque. SABIX*, 44, 119–129.
86. Darboux, G. (1878). Mémoire sur la théorie des coordonnées curvilignes et des systèmes orthogonaux. *Annales scientifiques de l'ENS*, 7, 227–260.
87. Love, A. E. H. (1927). *A Treatise on the Theory of Elasticity*. U.K., Cambridge: Cambridge University Press.
88. Vincensini, P. (1972). La géométrie différentielle au XIXème sciècle. *Scientia (Rivista di Scienza)*, 2–44.
89. Cartan, E. (1931). Géométrie Euclidienne et géométrie Riemannienne. *Scientia*, 49, 393-402. English translation in: (2007) Beyond Geometry - Classic papers from Riemann to Einstein. Dover Publications.
90. Riesz, F. (1910). Untersuchungen über Systeme integrierbarer Funktionen. *Mathematische Annalen*, LXIX, 449–496.
91. Hicks, J. (1966). Piet Hein Bestrides Art and Science. *Life*.
92. Gridgeman, N. T. (1970). Lamé ovals. *The Mathematical Gazette*, 54, 31.
93. Gielis, J. (1996). Wiskundige supervormen bij bamboes. *Newsletter of the Belgian Bamboo Society*, 13, 20–26.
94. Van Oystaeyen, F., Gielis, J., & Ceulemans, R. (1996). Mathematical aspects of plant modeling. *Scripta Botanica Belgica*, 13, 7–27.
95. Barr, A. H. (1981). Superquadrics and angle-preserving transformations. IEEE *Computer graphics and Applications*, 1(1), 11–23.
96. Jacklic, A., Leonardis, A. & Solina, F. (2000). *Segmentation and recovery of superquadrics*. The Netherlands, Dordrecht: Kluwer Academic Publisher.
97. Onaka, S. (2012). Superspheres: Intermediate shapes between spheres and polyhedra. *Symmetry*, 4(3), 336–343.
98. Onaka, S. (2016). Extended Superspheres for Shape Approximation of Near Polyhedral Nanoparticles and a Measure of the Degree of Polyhedrality. *Nanomaterials*, 6(2), 27.
99. Gielis, J., Shi, P., Ding, Y. (2016). Towards a geometrical theory of morphogenesis. Festschrifft, Festkolloqium Prof. Dr. Walter Liese, Reinbek, November 11, 2016.
100. Gielis J., Haesen S. & Verstraelen L. (2005). Universal shapes: from the supereggs of Piet Hein to the cosmic egg of George Lemaître. *Kragujevac Journal of Mathematics*, 28, 55–67.
101. Beirinckx, B., & Gielis, J. (2004). De Superformule. *Wiskunde en Onderwijs. Proceedings of VVWL Conf. in Oostende, July 1–2*.
102. Lenjou, K. (2005). *Krommen en Oppervlakken van Lamé en Gielis: van de formule van Pythagoras tot de superformule*. (Master's Thesis, University of Louvain, Departement of Mathematics).
103. Weisstein, E. W. (2003). Superellipse. From MathWorld. Retrieved from http://mathworld.wolfram.com/Superellipse.html
104. Weisstein, E. W. (2009). *CRC encyclopedia of mathematics*. U.S.A.: CRC Press.

105. Caratelli, D., Gielis, J., Tavkelidze, I., Ricci, P.E. (2013). The Dirichlet problem for the Laplace equation in supershaped annuli. *Boundary Value Problems,* 113. doi:10.1186/1687-2770-2013-113.
106. Weyl, H. (1952). *Symmetry.* Princeton University Press
107. Verstraelen, L. (2008). On Natural Geometric Symmetries. In: Atlantis Transactions on Geometry Vol 2. Dedicated to the memory of Katsumi Nomizu. Workshop "Differential Geometry and Submanifolds", Universidad de Murcia, Departamento de Matemáticas.
108. Gielis, J., Beirinckx, B. & Bastiaens, E. (2003). Superquadrics with rational and irrational symmetries. As cited in: Elber, G. & Shapiro, V (2003). Proceedings of the 8th ACM symposium on Solid Modeling and Applications, 262–265.
109. Caratelli, D., et al. (2010) *Fourier solution of the Dirichlet problem for the Laplace and Helmholtz equations in starlike domains.* Lecture Notes of Tbilisi International Centre of Mathematics and Informatics. Tbilisi University Press.
110. Gielis, J. (2004). Variational Superformula curves for 2D and 3D graphic arts. As cited in: Callaos, et al. (n.d.). Proceedings of 8th World Multi-Conference on Systemics, Cybernetics and Informatics, Volume V. Computer Science and Engineering, 119–124.
111. Bourke, P. http://paulbourke.net/geometry/supershape/Website
112. Gielis, J., Beirinckx, B. (2004). Superformula solutions for 3D graphic arts and CAD/CAM. SIGGRAPH 2004 sketches 0504.
113. Kiefer, A. (2005). *SupergraphX 3D – A Portfolio by Albert Kiefer.* The Netherlands, Venlo: SectorA.
114. Moltenbrey, K. (2004). Albert Kiefer – a portfolio. *Computer Graphics World* 27(12)
115. Tavkhelidze, I. (2000). On the Some Properties of One Class of Geometrical Figures. *TICMI,* 4, 51–55.
116. Tavkhelidze, I., Ricci, P.E. (2006). Classification of a wide set of Geometric figures, surfaces and lines (Trajectories), *Rendiconti Accademia Nazionale delle Scienze detta dei XL, Memorie di Matematica e Applicazioni,* 124o, vol. XXX, fasc. 1,. 191–212.
117. Tavkhelidze, I., Cassisa, C., Gielis, J. and Ricci, P.E. (2013). About Bulky Links, Generated by Generalized Mobius-Listing's bodies GML_3^n, *Rendiconti Lincei Mat. Appl.* 24, 11–38;
118. Tavkhelidze, I., et al. (2017). On a geometric model of bodies with "complex" configuration and some movements. *Atlantis Transactions in Geometry,* 2. Atlantis-Springer.
119. Goemans, W., & Van de Woestyne, I. (2014). Constant curvature twisted surfaces in 3-dimensional Euclidean and Minkowski 3-space. In Proceedings of the Conference RIGA 2014 Riemannian Geometry and Applications to Engineering and Economics. Romania, Bucharest: Publishing House of the University of Bucharest.
120. Goemans, W., & Van de Woestyne, I. (2016). Clelia curves, twisted surfaces and Plücker's conoid in Euclidean and Minkowski 3-space. *Recent Advances in the Geometry of Submanifolds:: Dedicated to the Memory of Franki Dillen (1963–2013),* 674, 59.
121. Goemans, W., & Van de Woestyne, I. (2015). Twisted Surfaces with Null Rotation Axis in Minkowski 3-Space. *Results in Mathematics,* 1–13.
122. von Goethe J.W. (1926). Goethes Morphologische Schriften. Ausgewahlt und eingeleitet von Wilhelm Troll. Eugen Diederichs, Jena.
123. Coen, E. S., & Carpenter, R. (1993). The Metamorphosis of Flowers. *The Plant Cell,* 5(10), 1175.
124. Darwin, C.R. (1859). *The origin of species by means of natural selection or the preservation of favoured races in the struggle for life.* John Murray, London.
125. Koenderink J.J. (2010). *Color for the Sciences.* Massachusetts: The MIT Press.
126. Gielis, J., Tavkelidze N I., Caratelli, D., Fougerolle, Y., Ricci, P.E., Gerats, T. (2012). Bulky knots and links generated by cutting Generalized Möbius Listing bodies and applications in the natural sciences. Royal Academy of Flanders. Proceedings of Math Art Summit 2012, Brussels.

127. Arnold, V.I., Oleinik, O. (1979). Topology of real algebraic manifolds. *Moscow Bulletin*, 34, 9–17.
128. Chacon, R. (2006). Modelling natural shapes with a simple nonlinear algorithm. *International Journal of Bifurcation and Chaos*, 16(8), 2365–2368.
129. Gielis, J., Natalini, P., Ricci, P.E. (2017). On generalized forms of the Gielis formula. In: Atlantis Transactions on Geometry Vol 2.
130. Graphic Arts Monthly (October 2004) Supplement: 2004 Intertech Technology Awards. S12
131. Gielis, J., Bastiaens, E., Krikken, T., Kiefer, A., de Blochouse, M. (2004). Variational Superformula curves for 2D and 3D graphic arts. SIGGRAPH, sketches 0442.
132. Gielis, J. & Gerats, T. (2004). A botanical perspective on plant shape modeling. As cited in: Savoie, M.J., et al. Proceedings of International Conference on Computing, Communications and Control Technologies, August 14–17, Austin, Texas. 265–272.
133. Wang, S., & Pan, J. Z. (2006). Integrating and querying parallel leaf shape descriptions. In International Semantic Web Conference, 668–681. Springer Berlin Heidelberg.
134. Gielis, J., Caratelli, D., Fougerolle, Y., Ricci, P.E., Gerats, T. (2011). Universal Natural Shapes: From unifying shape description to simple methods for shape analysis and boundary value problems. *PLOSOne*. D-11-01115R2 10.1371/journal.pone.0029324
135. Ricci, A. (1973). A constructive geometry for computer graphics. *The Computer Journal*, 16(2), 157–160.
136. Rvachev, V. (1974). *Methods of Logic Algebra in Mathematical Physics* (in Russian). Naukova Dumka, Kiev.
137. Rvachev, V.L., Rvachev, V.A. (1979). *Neklassicheskie metody teorii priblizhenij v kraevyh zadachah [Non-classical methods of approximation theory to boundary tasks]*. Kiev, Naukova dumka Publ.
138. Shapiro, V. (2007). Semi-analytic geometry with R-functions. *Acta Numerica*, 16, 239–303.
139. Lindenmayer, A. (1968). Mathematical models for cellular interactions in development II. Simple and branching filaments with two-sided inputs. *Journal of theoretical biology*, 18 (3), 300–315.
140. Prusinkiewicz, P., & Lindenmayer, A. (1989). *The algorithmic beauty of plants*. Germany, Berlin: Springer Verlag.
141. Meinhardt, H. (1998). *The algorithmic beauty of sea shells*. Germany, Berlin: Springer Verlag.
142. Mandelbrot, B.B. (1982). *The fractal geometry of nature*. New York: Henry Holt and Company.
143. Bosman, A.E. (1957). *Het wondere onderzoekingsveld der vlakke meetkunde*. The Netherlands, Breda: Parcival.
144. Fougerolle, Y., Truchetet, F., Gielis, J. (2017). Potential fields of self-intersecting Gielis curves for modeling and generalized blending techniques. Atlantis Transactions in Geometry, Volume 2.
145. Fougerolle, Y.D., Gribok, A., Foufou, S., Truchetet, F., Abidi, M.A. (2005). Boolean Operations with Implicit and Parametric Representation of Primitives Using R-Functions. *IEEE Transactions on Visualization and Computer Graphics*, 11(5), 529–539.
146. Fougerolle, Y.D., et al. (2006). Radial supershapes for solid modeling. *Journal of Computer Science and Technology*, 21 (2), 238–243.
147. Fougerolle, Y., Truchetet, F., Gielis, J. (2009). A new potential function for self intersecting Gielis curves with rational symmetries. In: International Conference on Computer Graphics Theory and Applications GRAPP'09, 90–95. INSTICC Press.
148. den Hoed, J., Peeters, S., Schaap, B., Smal, N., Verhoeven, L. (2014). SALMON Superformula And L-systems: Modelling Nature. Radboud Honours Academy, Faculty of Science, Radboud University Nijmegen.

149. Alexandrov, A. D., & Reshetnyak, Y. G. (1989). *General theory of irregular curves*, volume 29 of Mathematics and its Applications (Soviet Series).
150. Thom, R. (1991). *Prédire n'est pas expliquer. Entretiens avec Emile Noël.* Champs Flammarion.
151. Baez, J., & Stay, M. (2011). Physics, topology, logic and computation: a Rosetta Stone. In Coecke, B. (Ed.). (2011). New structures for physics (Vol. 813). Springer.
152. Boccaletti, D. (2001) Epicycles of the Greeks to Kepler's ellipse and the breakdown of the circle paradigm. Cosmology through time: ancient and modern cosmology in the meditteranean area. Monte Porzio Catone (Rome), Italy, June 18–20, 2001
153. Caratelli, D., Gielis, J., Ricci, P.E. (2011). Fourier-like solution of the Dirichlet problem for the Laplace equation in k-type Gielis domains. *Journal of Pure and Applied Mathematics: Advances and Applications* 5(2): 99–111.
154. Iwata, H., & Ukai, Y. (2002). SHAPE: a computer program package for quantitative evaluation of biological shapes based on elliptic Fourier descriptors. *Journal of Heredity*, 93(5), 384–385.
155. Chen, B.Y. (1984). *Total mean curvature and submanifolds and finite type curves*. Series in Pure Mathematics, 1. World Scientific.
156. Verstraelen, L. (1991). Curves and surfaces of finite Chen type. Geometry and Topology of Submanifolds, III, World Scientific, 304–311
157. Verstraelen, L. (2015). Psychology and Geometry I. On the geometry of the human kind. *Filomat,* 29(3), 545–552.
158. Gray, J. (2012*). Henri Poincaré: a scientific biography*. Princeton University Press.
159. Darrigol, O. (2005) The genesis of the Theory of Relativity. Séminaire Poincaré: 1–22
160. Gauss, C.F. (1818) Determinatio attractioni etc. *Werke,* Bd3, 333
161. Riemann, B. (1858). Über die Hypothesen, welche der Geometrie zu Grunde liegen. Habilitationsschrift, 1854, Abhandlungen der Königlichen Gesellschaft der Wissenschaften zu Göttingen, 13. English translations by Kingdon Clifford, W. (Nature Vol VII Nos; 183, 184, pp 14–17, 36, 37) or by Spivak, M. in: Beyond Geometry - Classic papers from Riemann to Einstein. Dover Publications 2007.
162. Finsler, P. (1918). *Über Kurven und Flachen in allgemeinen Raumen.* (Dissertation, Gottingen, 1918, Verlag Birkhauser Basel, 1951.)
163. Antonelli, P. L., & Miron, R. (2013). *Lagrange and Finsler Geometry. Applications to Physics and Biology*, 76. Springer Science & Business Media.
164. Arslan, K., Bulca, B., Bayram, B., Ozturk, G., Ugail, H. (2009) On Spherical Product Surfaces in E^3. International Conference on CyberWorlds, 2009, Bradford, West Yorkshire, UK September 07-September 11; pp. 132–137.
165. Haesen, S., Nistor, A., Verstraelen, L. (2012). On Growth and Form and Geometry I. *Kragujevac J.Mathematics*, 36(1), 5–25.
166. Grandi, G. (1713). Letter to Leibniz In: Leibnizens Mathematische Schriften Band IV, C.I. Gerhardt, 221–224.
167. Grandi, G. (1723). Florum Geometricorum Manipulus. *Phil. Trans.* 1722 32 355–371; doi:10.1098/rstl.1722.0070
168. Grandi, G. (1728) *Flores geometrici ex Rhodonearum et Cloeliarum curvarum descriptionibus resultantes.* Florence
169. Butzer, P., Jongmans, F. (1999) P.L. Chebyshev (1821–1894) A guide to his Life and Work. *Journal of Approximation Theory* 96: 111–138
170. Mason, J.C. & Handscomb, D.C. (2003) *Chebyshev polynomials*. Chapman and Hall/CRC
171. Gouzévitch, I., & Gouzévitch, D. (2009). Gabriel Lamé à Saint Pétersbourg (1820–1831). *Bulletin de la Sabix.* Société des amis de la Bibliothèque et de l'Histoire de l'École polytechnique, 44, 20–43.
172. Buschmann, R.G. (1963). Fibonacci Numbers, Chebyshev Polynomials, Generalizations and Difference Equations. *The Fibonacci Quarterly*, 1(4), 1–7.

173. Darboux, J. (1904). A study of the development of geometric methods. *Popular science monthly* Vol LXVI: 412–434. The Science Press, New York
174. Green, P. B. (1999). Expression of pattern in plants: combining molecular and calculus-based biophysical paradigms. *American Journal of Botany*, 86(8), 1059–1076.
175. Fourier, J. (1952). Analytical theory of heat. Great Books of the Western World, Encyclopedia Brittanica.
176. Natalini, P., Patrizi, R., & Ricci, P. E. (2008). The Dirichlet problem for the Laplace equation in a starlike domain of a Riemann surface. *Numerical Algorithms*, 49(1–4), 299–313.
177. Natalini, P., Patrizi, R., & Ricci, P. E. (2009). Heat problems for a starlike shaped plate. *Applied Mathematics and Computation*, 215(2), 495–502.
178. Caratelli, D., Natalini, P., Ricci, P.E. (2010). Fourier solution of the wave equation for a starlike shaped vibrating membrane. *Computers and Mathematics with Applications*, 59, 176–184.
179. Caratelli, D., Natalini, P., Ricci, P.E., Yarovoy, A. (2010). The Neumann problem for the Helmholtz equation in a starlike planar domain. *Applied Mathematics and Computation*, 216(2), 556–564.
180. Caratelli, et al. (2009). Fourier solution of the Dirichlet problem for the Laplace and Helmholtz equations in starlike domains. Lecture Notes of TICMI, 10.
181. Caratelli, D., Gielis, J., Ricci, P.E., Tavkelidze, I. (2013). *Boundary Value Problems*. DOI: 10.1186/10.1186/1687-2770-2013-253.
182. Fournier, M., P. Rogier, E. Costes, and M. Jaeger. 1993. Modélisation mécanique des vibrations propres d'un arbre soumis aux vents, en fonction de sa morphologie. *Annales des Sciences Forestières* 50: 401–412.
183. Rodriguez, M., De Langre, E., Moulia, B. (2008). A scaling law for the effects of architecture and allometry on tree vibration modes suggests a biological tuning to modal compartmentalization. *Am. J. Botany* 95(12):1523–1537.
184. Niklas, K. J. (1985). The aerodynamics of wind pollination. *The Botanical Review*, 51(3), 328–386.
185. Leissa, A.W. (1969). *Vibrations of Plates*. NASA SP-160.
186. Turing, A. (1952). The Chemical basis of morphology. *Philosophical Transactions of the Royal Society of London*, 237(641), 37–72.
187. Gielis, J., Caratelli, D., Fougerolle, Y., Ricci, P.E, Gerats, T. (2017) A biogeometrical model for corolla fusion in asclepiad flowers. Atlantis Transactions in Geometry Vol 2.
188. Vekua (2013). On Metaharmonic functions. *Lecture Notes of TICMI 14*, Tbilisi University Press.
189. Rvachev, V.L., Sheiko, T.I., Shapiro, V., Tsukanov, I. (1999). On completeness of RFM solutions structures. SAL-2.
190. Alexandrov, A.D. (1963). Curves and surfaces. In: Lavrent'ev, A. D., Aleksandrov, A. D., . Kolmogorov: Mathematics: its content, methods and meaning. Cambridge: M.I.T. Press, MA.
191. Kühnel, W. (2008). *Differentialgeometrie*. Vieweg Studium - Aufbaukurs Mathematik.
192. Verstraelen, L. (2013). Geometry of submanifolds I. The first Casorati Curvatures indicatrices *Kragujevac Journal of Mathematics*, 37(1), 5–23.
193. Catalan, E. (1891). Sur le courbure des surfaces. Lettre adressée a M. Casorati. *Acta Mathematica*.
194. Koenderink, J.J. (1990). *Solid Shape*. Massachusetts: The MIT Press.
195. Ferrandez, A. (2017). Some variational problems on curves and applications. Atlantis Transactions in Geometry, 1, 199–222.
196. Chen, B.Y. (2003). Riemannian DNA, Inequalities and their applications. *Tamkang Journal of Science and Engineering*, 3(3), 123–130.
197. Koiso, M., & Palmer, B. (2007). Anisotropic capillary surfaces. As cited in: Dillen, F., Simon, U., Vrancken, L. (2007). Symposium on the Differential Geometry of Submanifolds, Valenciennes, 185–196.

198. Koiso, M., & Palmer, B. (2008). Rolling Construction for Anisotropic Delaunay Surfaces. *Pacific J. of Math.*, 2, 345–378.
199. Haesen, S., Sebekovic, A. & Verstraelen, L. (2003). Relations between intrinsic and extrinsic curvatures. *Kragujevac Journal of Mathematics*, 25, 139–145.
200. Hopf, H. (1983). *Differential geometry in the large*. Germany, Berlin: Springer Verlag.
201. Cartan, E (1972). *Exposés de géométrie*. Hermann, Paris
202. Von Goethe, J.W. (1831). Über die Spiraltendenz der Vegetation. Manuscript, Weimar.
203. Chen, B.-Y. (2017). Topics in differential geometry associated with position vector fiels in Euclidean submanifolds. *Arab Journal of Mathematical Sciences* 23: 1–17.
204. Fougerolle, Y.D., Gribok, A., Foufou, S., Truchetet, F., Abidi, M.A. (2006). Supershape recovery from 3D data sets. As cited in: (2006). Proceedings of the International Conference on Image Processing ICIP'06, Atlanta, 2193–2196.
205. Fougerolle, Y., Gielis, J., Truchetet, F. (2013). A robust evolutionary algorithm for the recovery of rational Gielis curves, *Pattern Recognition*, vol 46(8), 2078–2091.
206. Jean, R. V. (1994). *Phyllotaxis: a systemic study in plant morphogenesis*. U.K., Cambridge: Cambridge University Press.
207. McLellan, T. (1993). The roles of heterochrony and heteroblasty in the diversification of leaf shapes in Begonia dreigei (Begoniaceae). *American Journal of Botany*, 80, 796–804.
208. Van Droogenbroeck, B., Breyne, P., Goetghebeur, P., Romeijn-Peeters, E., Kyndt, T., & Gheysen, G. (2002). AFLP analysis of genetic relationships among papaya and its wild relatives (Caricaceae) from Ecuador. *Theoretical and Applied Genetics*, 105(2–3), 289–297.
209. Shi, P-J., et al. (2015) Comparison of dwarf bamboo (Indocalamus sp.) leaf parameters to determine relationship between spatial density of plants and total leaf area per plant. *Ecology and Evolution*. doi: 10.1002/ece3.1728
210. Lin, S., et al. (2016). A geometrical model for testing bilateral symmetry of bamboo leaf with a simplified Gielis equation. *Ecology and Evolution*, 6(19), 6798–6806.
211. Brandis, D. (1907). Remarks on the Structure of Bamboo Leaves. *Transactions of the Linnean Society of London*. 2nd Series: Botany, 7(5), 69–92.
212. Shi, P-J., et al. (2015) Capturing spiral growth of conifers using superellipse to model tree-ring geometric shape. *Front. Plant Sci.*, 6, 856. doi:10.3389/fpls.2015.00856
213. Darwin, C. (1865). On the movements and habits of climbing plants. *Journal of the Linnean Society of London, Botany*, 9(33–34), 1–118.
214. Liese, W., & Ammer, U. (1962). Anatomische Untersuchungen an extrem drehwüchsigem Kiefernholz. *Holz als Roh-und Werkstoff*, 20(9), 339–346.
215. Wei, Q., et al. (2016). Exploring key cellular processes and candidate genes regulating the primary thickening growth of Moso underground shoots. *New Phytologist*. doi:10.1111/nph.14284
216. Faisal, T.R., Abad, E.M.K., Hristozov, N., Pasini, D. (2010). The Impact of Tissue Morphology, Cross-Section and Turgor Pressure on the Mechanical Properties of the Leaf Petiole in Plants. *Journal of Bionic Engineering*, 7(1), S11–S23.
217. Faisal, T. R., Hristozov, N., Western, T. L., Rey, A., & Pasini, D. (2016). The twist-to-bend compliance of the Rheum rhabarbarum petiole: integrated computations and experiments. *Computer Methods in Biomechanics and Biomedical Engineering*, 1–12.
218. Beirinckx, B. (1997). *Supervormen, wortels en bamboe*. (Master's thesis, Hoboken, Belgium).
219. Beirinckx, B. (2003). Cirkel is niet optimaal. Genicap Corporation, Internal Report.
220. Niklas, K.J. (1992). *Plant biomechanics. An engineering approach to to Plant Form and Function*. Chicago: University of Chicago Press.
221. Niklas, K.J., Spatz, H.C (2012). *Plant Physics*. Chicago: University of Chicago Press.
222. Caratelli, D., Mescia, L., Bia, P., & Stukach, O. V. (2016). Fractional-calculus-based FDTD algorithm for ultrawideband electromagnetic characterization of arbitrary dispersive dielectric materials. *IEEE Transactions on Antennas and Propagation*, 64(8), 3533–3544.

223. Rodrigo, J. A., & Alieva, T. (2016). Polymorphic beams and Nature inspired circuits for optical current. *Scientific Reports*, 6.
224. Silva, P. F., Freire, R. C. S., Serres, A. J. R., Silva, P. D. F., & Silva, J. C. (2016). Wearable textile bioinspired antenna for 2G, 3G, and 4G systems. *Microwave and Optical Technology Letters*, 58(12), 2818–2823.
225. Arditti, J., Elliott, J., Kitching, I. J., & Wasserthal, L. T. (2012). 'Good Heavens what insect can suck it'–Charles Darwin, Angraecum sesquipedale and Xanthopan morganii praedicta. *Botanical Journal of the Linnean Society*, 169(3), 403–432.
226. Wallace, A. R. (1867). Creation by Law. *Quarterly Journal of Science*, 4, 471–488.
227. Graham, A. W., et al. (2012). LEDA 074886: A remarkable rectangular-looking galaxy. *The Astrophysical Journal*, 750(2), 121.
228. Walsby, A. E. (1980). A square bacterium. *Nature*, 283, 69–71.
229. Gaffney, L. P., et al. (2013). Studies of pear-shaped nuclei using accelerated radioactive beams. *Nature*, 497(7448), 199–204.
230. Becker, T.R. (2000). Taxonomy, evolutionary History and Distribution of the middle to late Famennian Wocklumeriina (Ammonoidea, Clymeniida). *Mitt. Mus. Nat.kd. Berl., Geowiss. Reihe*, 3, 27–75.
231. Korn, D., Klug, C., & Mapes, R. H. (2005). The Lazarus ammonoid family Goniatitidae, the tetrangularly coiled Entogonitidae, and Mississippian biogeography. *Journal of Paleontology*, 79(02), 356–365.
232. Ebbighausen, V., & Korn, D. (2007). Conch geometry and ontogenetic trajectories in the triangularly coiled Late Devonian ammonoid Wocklumeria and related genera. *Neues Jahrbuch für Geologie und Paläontologie-Abhandlungen*, 244(1), 9–41.
233. Janner, A. (1997). De nive sexangula stellata. *Acta Crystallographica Section A: Foundations of Crystallography*, 53(5), 615–631.
234. Janner, A. (2001). DNA enclosing forms from scaled growth forms of snow crystals. *Crystal Engineering*, 4(2), 119–129.
235. Libbrecht, K. G. (2005). The physics of snow crystals. *Reports on progress in physics*, 68(4), 855.
236. Viau, C. (2012). *Software for CAMC surfaces.*
237. Telford, M., & Mooi, R. (1987). The art of standing still. *New Scientist*, 114(1556), 30–35.
238. Darwin, C. (1862). *On the various contrivances by which British and foreign orchids are fertilised by insects: and on the good effects of intercrossing.* U.K., London: J. Murray.
239. Harder, L. D., & Johnson, S. D. (2008). Function and evolution of aggregated pollen in angiosperms. *International Journal of Plant Sciences,* 169(1), 59–78.
240. Endress, P. K. (2015). Development and evolution of extreme synorganization in angiosperm flowers and diversity: a comparison of Apocynaceae and Orchidaceae. *Annals of botany*, mcv119.
241. Schwartz-Sommer, Z., Huijser, P., Nacken, W., Saedler, H., Sommer, H. (1990). Genetic control of flower development by homeotic genes in *Antirrhinum majus*. *Science,* 250, 931–936.
242. Coen, E.S., & Meyerowitz, E.M. (1990). The war of the whorls: genetic interactions controlling flower development. *Nature,* 353, 31–37.
243. Rijpkema, A.S., Vandenbussche, M., Koes, R., Heijmans, K., & Gerats, T. (2010). Variations on a theme: changes in the floral ABCs in angiosperms. *Seminars in Cell & Developmental Biology*, 21(1),100–107.
244. Causier, B., Schwartz-Sommer, Z., & Davies, B. (2010). Floral organ identity: 20 years of ABCs. *Seminars in Cell & Developmental Biology*, 21, 73–79.
245. Heijmans, K., et al. (2012). Redefining C and D in the Petunia ABC. *The Plant Cell*, 24(6), 2305–2317.
246. Endress, P.K. (2001). Origins of flower morphology. *J.Exp.Zool.(Mol.Dev.Evol.),* 291, 105–115.

247. Ronse-De Craene, L., Smets, E. (1993). The distribution and systematic relevance of the androecial character polymery. *Botanical journal of the Linnean Society*, 113, 285–350.
248. Ronse-De Craene, L. (2011). *Floral Diagrams*. U.K., Cambridge: Cambridge University Press.
249. Endress, P.K. (2011). Evolutionary diversification of the flowers in angiosperms. *American Journal of Botany*, 98(3), 370–396.
250. Vandenbussche, M., Theissen, G., Van de Peer, Y., & Gerats, T. (2003). Structural diversification and neo-functionalization during floral MADS-box gene evolution by C-terminal frameshift mutations. *Nucleic Acids Research*, 31(15), 4401–4409.
251. Gerats, T., & Vandenbussche, M. (2005). A model system for comparative research: Petunia. *Trends in plant science*, 10(5), 251–256.
252. Vandenbussche, M., Horstman, A., Zethof, J., Koes, R., Rijpkema, A., & Gerats, T. (2009). Differential recruitment of WOX transcription factors for lateral development and organ fusion in Petunia and Arabidopsis. *Plant Cell*, 21, 2269–2283.
253. Alberts, F., & Meve, U. (2002). *Illustrated Handbook of Succulent Plants: Asclepiadaceae*. Springer.
254. Vereecken, N.J., McNeil, J.N. (2010). Cheaters and liars: chemical mimicry at its finest. *Can. J. Zool*, 88, 725–752. doi:10.1139/Z10-040
255. Endress, P.K. (1994). *Diversity and evolutionary biology of tropical flowers*. Cambridge University Press.
256. Costanzo, E., Trehin, C., & Vandenbussche, M. (2014). The role of WOX genes in flower development. *Annals of botany*, 114(7), 1545–53.
257. Vandenbussche, M., Chambrier, P., Bento, S. R., & Morel, P. (2016). Petunia, Your Next Supermodel? *Frontiers in plant science*, 7.
258. Zhong, J., & Preston, J.C. (2015). Bridging the gaps: evolution and development of perianth fusion. *New Phytologist*, 208.2, 330–335.
259. Niu, L., et al. (2015). LOOSE FLOWER, a WUSCHEL-like Homeobox gene, is required for lateral fusion of floral organs in Medicago truncatula. *The Plant Journal*, 81(3), 480–492.
260. Endress, P. K. (2010). Synorganisation without organ fusion in the flowers of Geranium robertianum (Geraniaceae) and its not so trivial obdiplostemony. *Annals of botany*, mcq171.
261. Angiosperm Phylogeny Group (2016). An update of the Angiosperm Phylogeny Group classification for the orders and families of flowering plants: APG IV. *Botanical Journal of the Linnean Society* 181(1): 1–20
262. Reyes, E., Sauquet, H., & Nadot, S. (2016). Perianth symmetry changed at least 199 times in angiosperm evolution. *Taxon*, 65.5, 945–964.
263. Bombarely, A., et al. (2016). Insight into the evolution of the Solanaceae from the parental genomes of Petunia hybrida. *Nature plants, 2,* 16074.
264. Thom, R. (1980). Halte au hasard, silence au bruit. *Le débat*, 3, 119–132.
265. Niklas, K. J., & Kutschera, U. (2016). From Goethe's plant archetype via Haeckel's biogenetic law to plant evo-devo. *Theory in Biosciences*, 1–9.
266. Van der Waerden, B.L. (1972). Über die Wechselwirkung zwischen Mathematik und Physik. *Elemente der Mathematik*. 33–41